ok 2

Maths Frameworking

3rd edition

Peter Derych, Kevin Evans,
Keith Gordon, Michael Kent,
Trevor Senior, Brian Speed

William Collins' dream of knowledge for all began with the publication of his first book in 1819. A self-educated mill worker, he not only enriched millions of lives, but also founded a flourishing publishing house. Today, staying true to this spirit, Collins books are packed with inspiration, innovation and practical expertise. They place you at the centre of a world of possibility and give you exactly what you need to explore it.

Collins. Freedom to teach.

Published by Collins
An imprint of HarperCollins Publishers
The News Building
1 London Bridge Street
London
SE1 9GF

Acknowledgements
The publishers wish to thank the following for permission to reproduce photographs. Every effort has been made to trace copyright holders and to obtain their permission for the use of copyright materials. The publishers will gladly receive any information enabling them to rectify any error or omission at the first opportunity.

Cover gyn9037/Shutterstock.

Browse the complete Collins catalogue at
www.collins.co.uk

10 9 8

ISBN-13 978-0-00-753764-8

British Library Cataloguing in Publication Data
A Catalogue record for this publication is available from the British Library.

Written by Peter Derych, Kevin Evans, Keith Gordon, Michael Kent, Trevor Senior, Brian Speed
Commissioned by Katie Sergeant
Project managed by Elektra Media Ltd
Development edited and copy-edited by Susan Gardner
Edited by Helen Marsden
Proofread by Joanna Shock
Illustrations by Ann Paganuzzi
Typeset by Jouve India Private Limited
Page and cover design by Angela English

Printed and bound by Grafica Veneta S.p.A. Italy.

Contents

How to use this book 5

1 Working with numbers 6

1.1 Multiplying and dividing negative numbers 6
1.2 Highest common factors (HCF) 7
1.3 Lowest common multiples (LCM) 8
1.4 Powers and roots 9
1.5 Prime factors 10

2 Geometry 11

2.1 Angles in parallel lines 11
2.2 The geometric properties of quadrilaterals 13
2.3 Rotations 14
2.4 Translations 17
2.5 Constructions 19

3 Probability 21

3.1 Probability scales 21
3.2 Mutually exclusive events 22
3.3 Sample space diagrams 24
3.4 Experimental probability 26

4 Percentages 28

4.1 Calculating percentages 28
4.2 Percentage increase and decrease 29
4.3 Percentage change 31

5 Sequences 32

5.1 Using flow diagrams to generate sequences 32
5.2 Using the nth term of a sequence 33
5.3 Finding the nth term of a sequence 34
5.4 The Fibonacci sequence 35

6 Area of 2D and 3D shapes 37

6.1 Area of a triangle 37
6.2 Area of a parallelogram 39
6.3 Area of a trapezium 41
6.4 Surface area of cubes and cuboids 43

7 Graphs 45

7.1 Graphs from linear equations 45
7.2 Gradient of a straight line 46
7.3 Graphs from simple quadratic equations 48
7.4 Real-life graphs 49

8 Simplifying numbers 52

8.1 Powers of 10 52
8.2 Large numbers and rounding 53
8.3 Significant figures 54
8.4 Standard form with large numbers 55
8.5 Multiplying with numbers in standard form 57

9 Interpreting data 58

9.1 Pie charts 58
9.2 Creating pie charts 60
9.3 Scatter graphs and correlation 62
9.4 Creating scatter graphs 64

10 Algebra 66

10.1 Algebraic notation 66
10.2 Like terms 67
10.3 Expanding brackets 68
10.4 Using algebraic expressions 69
10.5 Using index notation 71

11 Congruence and scaling 73

11.1 Congruent shapes 73
11.2 Enlargements 76
11.3 Shape and ratio 78
11.4 Scales 80

12 Fractions and decimals 84

12.1 Adding and subtracting fractions 84
12.2 Multiplying with fractions and integers 85
12.3 Dividing with fractions and integers 87
12.4 Multiplication with large and small numbers 88
12.5 Division with large and small numbers 89

13 Proportion 91

13.1 Direct proportion 91
13.2 Graphs and direct proportion 92

| **13.3** | Inverse proportion | 94 |
| **13.4** | Comparing direct proportion and inverse proportion | 96 |

14 Circles — 99

14.1	The circle and its parts	99
14.2	Formula for the circumference of a circle	100
14.3	Formula for the area of a circle	102

15 Equations and formulae — 105

15.1	Equations with brackets	105
15.2	Equations with the variable on both sides	106
15.3	More complex equations	108
15.4	Rearranging formulae	109

16 Comparing data — 111

16.1	Grouped frequency tables	111
16.2	Drawing frequency diagrams	113
16.3	Comparing range and averages	115
16.4	Which average to use?	117

How to use this book

Welcome to *Maths Frameworking 3rd edition Homework Book 2*

Maths Frameworking Homework Book 2 accompanies Pupil Books 2.1, 2.2 and 2.3 and has hundreds of practice questions at different levels to help you consolidate what you have learned in class. Key features enable easy navigation through the book: indicators help you find the right questions for your level and question type icons help you find and practise key skills.

These are the key features:

Numbered topics match the Pupil Books so you can find the right sections easily.

The **level of difficulty** of the questions corresponds to the three different year 2 Pupil Books:

2.1   **2.2**   **2.3**

Practise your **problem solving, mathematical reasoning** and **financial skills** with highlighted questions.

(PS) (MR) (FS)

Challenge yourself with extended **Brainteaser** activities.

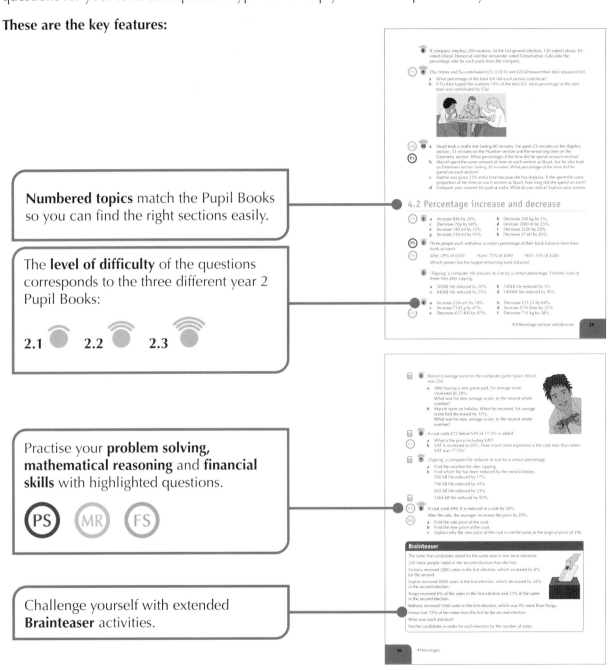

1 Working with numbers

1.1 Multiplying and dividing negative numbers

1 Calculate the following.

a 5 × −3 **b** −3 × 10 **c** −8 × −5
d −2 × −9 **e** 14 × −3 **f** 4 × −3 × 2
g 2 × −3 × −7 **h** −10 × −6 × −10

2 Calculate the following.

a 16 ÷ −2 **b** −6 ÷ −6 **c** −9 ÷ 3
d −12 ÷ −3 **e** −100 ÷ 10 **f** 20 ÷ −2 ÷ −5
g −36 ÷ −3 ÷ −4 **h** −40 ÷ 5 ÷ −2

PS **3** **a** There are eight calculations that multiply whole numbers to give the answer −6. How many can you find?
b Write down five calculations involving division that give the answer −5.

4 Find the missing number in each calculation.

a 2 × −5 = ☐ **b** −2 × ☐ = −16 **c** 3 × ☐ = −15
d −4 × −7 = ☐ **e** ☐ × −6 = 36 **f** −3 × ☐ = −27
g 60 ÷ ☐ = −30 **h** −45 ÷ 3 = ☐ **i** ☐ ÷ −5 = −6
j ☐ ÷ 6 = −7 **k** ☐ ÷ −6 = −7 **l** 28 ÷ ☐ = −4

5 **a** Copy and complete this grid.

×	4	−5	6	−7
−2				
3				
−8				
9				

b Copy and complete this grid.

×	−1		−6	
−3		−24		
	−5			20
7				
			54	

6 Calculate the following.

    a $-3 \times -3 = 9$ ✓   **b** $(-5)^2 = 25$ ✓   **c** $(-8)^2 = 64$ ✓

    d $49 \div -7 = -7$ ✓   **e** $-36 \div 6 = -6$ ✓

 7 Calculate the following.

    a $2 \times -6 \div -4 = 3$ ✓   **b** $4 \times 6 \div -3 = -8$ ✓   **c** $-8 \times -7 \div 2 = 28$ ✓

    d $20 \div -5 \times 9 = -36$ ✓   **e** $-30 \div 5 \times -7 = 42$ ✓   **f** $-2 \times -2 \times 2 = 8$ ✓

    g $4 \times -4 \times 4 = -64$ ✓   **h** $-3 \times -3 \times -3 = -27$ ✓   **i** $(-5)^3 = -125$ ✓

 8 Find the missing number in each calculation.

    a $-2 \times \boxed{-2} \times 3 = 12$ ✓   **b** $-5 \times \square \times -2 = 30$ **c** $\square \times -4 \times -2 = -24$

    d $-20 \div \square \times 3 = 12$ **e** $5 \times 4 \div \square = -10$ **f** $36 \div \square \div -3 = 3$

Brainteaser

Give two answers for each of the following.

a $\sqrt{16}$ **b** $\sqrt{25}$ **c** $\sqrt{100}$

1.2 Highest common factors (HCF)

 1 **a** Write out all the factors of the following numbers.

       i 12   **ii** 15   **iii** 20   **iv** 30

    b Which two numbers are always factors of any number?

 2 A PE teacher wants to split 30 pupils into teams for various activities.

    a How many pupils will be in each team if there are **i** 2 teams **ii** 5 teams?

    b How many teams will there be if each team has **i** 3 pupils **ii** 6 pupils?

    c Why would it not be easy to have 4 teams?

 3 **a** What are the four common factors of 12 and 30?

    b Which is the highest common factor (HCF) of 12 and 30?

 4 Use your answers to question **1** to find the HCF of the following pairs of numbers.

    a 12 and 15   **b** 12 and 20   **c** 20 and 30   **d** 15 and 20

  5 Two teachers take 48 pupils on a trip. They want to divide pupils into equal-sized groups. What possible sized groups could they use?

6 Write down all the common factors of the following pairs of numbers.

    a 12 and 18   **b** 24 and 32   **c** 15 and 45

    d 30 and 75   **e** 28 and 42   **f** 36 and 48

 7 Find the largest number that can divide into each of these pairs of numbers.

    a 20 and 60   **b** 25 and 75   **c** 32 and 80

    d 40 and 120   **e** 60 and 75   **f** 36 and 180

8 Find the HCF of the following pairs of numbers.

    a 14 and 35   **b** 8 and 20   **c** 12 and 30   **d** 15 and 24

 9 Simplify each of the following fractions.

    a $\dfrac{24}{30}$ **b** $\dfrac{15}{18}$ **c** $\dfrac{10}{25}$ **d** $\dfrac{21}{45}$

    e $\dfrac{20}{64}$ **f** $\dfrac{28}{42}$ **g** $\dfrac{104}{160}$ **h** $\dfrac{63}{81}$

 10 a There are three common factors of 8, 20 and 32. Work them out.

    b Which is the highest common factor of 8, 20 and 32?

1.3 Lowest common multiples (LCM)

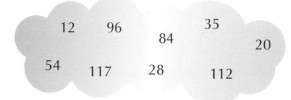

12   96   84   35   20

54   117   28   112

1 Which of the numbers in the bubble above are multiples of:

    a 4   **b** 7   **c** 9   **d** 12?

 2 Write the first 10 multiples of the following numbers.

    a 3   **b** 5   **c** 8   **d** 12

3 a Write down the first three common multiples of 3 and 5.

    b Which is the lowest?

 **4** Use your answers to question **2** to find the LCM of the following pairs of numbers.

    a 3 and 8   **b** 5 and 12   **c** 5 and 8   **d** 8 and 12

5 Find the LCM of the following pairs of numbers.

    a 4 and 8   **b** 6 and 10   **c** 7 and 8   **d** 9 and 12

 6 Find the LCM of the following.

    a 2, 3 and 5   **b** 3, 4 and 5   **c** 5, 6 and 8   **d** 6, 9 and 15

    e 3, 8 and 12   **f** 5, 12 and 16   **g** 7, 9 and 12   **h** 3, 7 and 11

 7 What are the common denominators of these fractions?

    a $\frac{2}{3}$ and $\frac{7}{15}$ **b** $\frac{3}{4}$ and $\frac{5}{6}$ **c** $\frac{3}{5}$ and $\frac{4}{9}$ **d** $\frac{1}{2}$, $\frac{2}{5}$ and $\frac{3}{8}$

 8 Add together each set of fractions in question **7**.

Brainteaser

a Two athletes run around an indoor running track. One completes a lap in 24 seconds, the other in 30 seconds. Assuming they continue to run at the same pace for successive laps, how many laps will it be before the faster runner passes the slower one?

b The same athletes take 48 seconds and 56 seconds respectively on an outdoor track. How many laps will it take for the faster one to catch the slower one this time?

c Can you see a clue that would help you to get the answer to parts **a** and **b** quickly?

1.4 Powers and roots

 1 Work out the cubes of these numbers without a calculator.

    a 4   **b** 5   **c** 10

 2 Use your calculator to find the following.

    a 20^2 **b** 15^3 **c** 25^3 **d** 6.6^2 **e** 4.2^3 **f** 7.3^3

 3 Use your calculator to find the following.

    a 2^6 **b** 4^5 **c** 8^4 **d** 6^6 **e** 3^8

 4 Write down the positive values of the following roots.

    a $\sqrt{16}$ **b** $\sqrt{81}$ **c** $\sqrt[3]{64}$ **d** $\sqrt[3]{343}$

5 Find two values of x that make each equation true.

    a $x^2 = 49$ **b** $x^2 = 100$ **c** $x^2 = 225$ **d** $x^2 = 1.44$

 6 **a** Use your calculator to find the following.

      i 0.4^2 **ii** 0.5^2

    b Work out the answer to 0.6^2 in your head. Now check the answer with your calculator.

    c Work out the answer to 0.3^2 in your head. Again check the answer with your calculator.

    d Copy and complete the table below.

Number	0.1	0.2	0.3	0.4	0.5	0.6	0.7	0.8	0.9	1
Square										

  7 Choose any number between 0 and 1, for example, 0.4.

    a Calculate increasing powers of your number, for example, 0.4^2, 0.4^3, 0.4^4.

    b What do you notice about the sizes of your answers as the power increases? Explain why this happens.

 8 **a** Use your calculator to find the following.

      i 0.5^3 **ii** 0.6^3

    b Copy and complete the table below.

Number	0.1	0.2	0.3	0.4	0.5	0.6	0.7	0.8	0.9	1
Cube										

1.5 Prime factors

1 Calculate these products of prime factors.

    a $2 \times 5 \times 5$ **b** $2^3 \times 3$ **c** $2 \times 3 \times 7^2$

2 Use a prime factor tree to write each of the following numbers as a product of its prime factors.

    a 6   **b** 18   **c** 32   **d** 70   **e** 36

3 Use the division method to write each of the following numbers as a product of its prime factors.

    a 14   **b** 45   **c** 96   **d** 130   **e** 200

 4 Find the prime factors of all the numbers from 21 to 30.

 5 **a** Which numbers in question **4** only have one prime factor?

    b What special name is given to these numbers?

    c What is the next number after 30 with only 1 prime factor?

6 The prime factors of 60 are $2 \times 2 \times 3 \times 5 = 2^2 \times 3 \times 5$.

    a Write down the prime factors of 120 in index form.

    b Write down the prime factors of 180 in index form.

    c Write down the prime factors of 600 in index form.

2 Geometry

2.1 Angles in parallel lines

1 Which of these sets of lines are parallel?

a 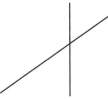 **b**  **c**

d 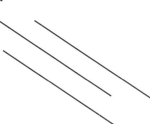 **e**  **f**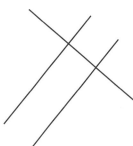

2 Copy and complete these sentences.

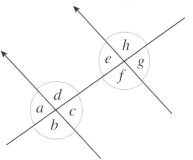

    a f and ___ are alternate angles.
    b c and ___ are corresponding angles.
    c ___ and a are corresponding angles.
    d c and e are _____ angles
    e d and h are _____ angles.

3 **a** Write down three pairs of alternate angles.
    b Write down three pairs of corresponding angles.

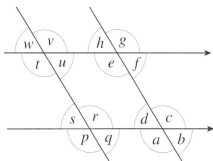

MR **4** Which diagram is the odd one out?

Give a reason for your answer.

a  **b**  **c**

5 Calculate the size of each unknown angle.

State whether it is an alternate angle or a corresponding angle.

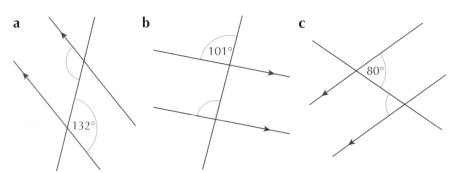

a   **b**   **c**

132°   101°   80°

MR **6** Copy each of the following diagrams.

Calculate the sizes of all the angles and mark them on your diagram.

Explain how you worked out your answers.

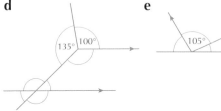

a   **b**   **c**

40°   125°   32°   97°   43°

d   **e**

135° 100°   105°   53°

MR **7** Use alternate and/or corresponding angles to explain why opposite angles in a rhombus are the same size.

MR **8** Copy the following diagram.

Calculate the sizes of all the angles and mark them on your diagram.

Explain how you worked out your answers and why the two triangles are the same shape.

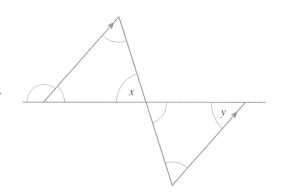

x

y

2.2 The geometric properties of quadrilaterals

1   Copy this table and, in each column, write the names of all possible quadrilaterals that could fit the description.

Two pairs of equal angles	Rotational symmetry of order 4	Exactly one line of symmetry	Exactly two lines of symmetry	Exactly two right angles	Exactly four equal sides

(**Warning:** A quadrilateral could be in more than one column!)

2   **a**   Some quadrilaterals have two pairs of equal angles. Which are they?
    b   Some quadrilaterals have two pairs of equal sides. Which are they?

3   **a**   A quadrilateral has exactly two equal angles.

        What type of quadrilateral could it be?

    b   A quadrilateral has exactly three equal angles.

        What type of quadrilateral could it be?

MR 4   Which quadrilaterals have, or could have, diagonals that intersect at right angles? Illustrate your answer with drawings.

5   State all the quadrilaterals which could have:

    a   at least two right angles
    b   no lines of symmetry
    c   no parallel sides
    d   rotational symmetry of order 2 or more.

MR 6   The diagonals of a quadrilateral divide it into four triangles.

    Which quadrilaterals contain at least one isosceles triangle?

    Illustrate your answer with drawings.

7   Identify the quadrilateral from its angles (there may be more than one answer).
    a   90°, 90°, 90°, 90°
    b   72°, 72°, 108°, 108°
    c   72°, 90°, 90°, 108°
    d   30°, 30°, 30°, 270°
    e   75°, 85°, 95°, 105°

2.3 Rotations

1 Copy each of the triangles below and draw the image after each one has been rotated about the point marked X through the angle indicated. Use tracing paper to help.

90° clockwise   180°   90° anticlockwise

2 Copy each of the shapes below onto a square grid. Draw the image after each one has been rotated about the point marked X through the angle indicated. Use tracing paper to help.

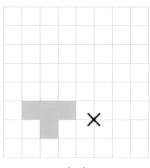   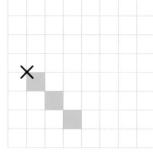   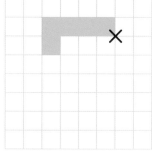

90° clockwise   90° anticlockwise   90° anticlockwise

3 Copy each of these isosceles triangles onto a coordinate grid, with axes for x and y from −5 to 5.

    a   Draw the image A′B′C′ of each one after it has been rotated about the origin, O, through the angle and direction indicated.
    b   Write down the coordinates of each object.
    c   Write down the coordinates of each image.

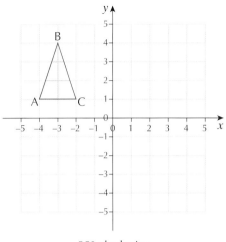

90° clockwise

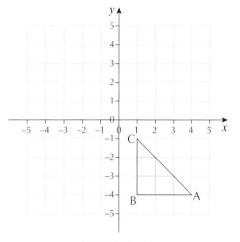

180° clockwise

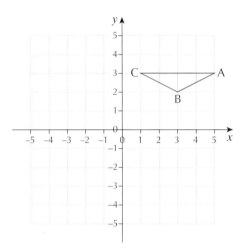

90° anticlockwise

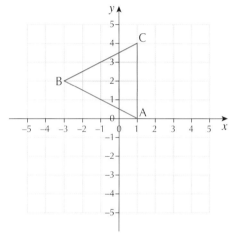

180° anticlockwise

4 Copy the trapezium onto a coordinate grid, with axes for x and y from 0 to 10.

a Rotate the trapezium ABCD through 90° anticlockwise about the point (4, 6) to give the image A′B′C′D′.

b Write down the coordinates of A′, B′, C′ and D′.

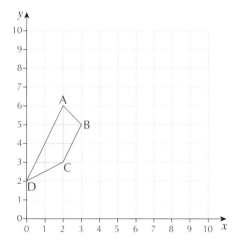

5 Copy the pentagon onto a coordinate grid, with axes for x and y from 0 to 10.

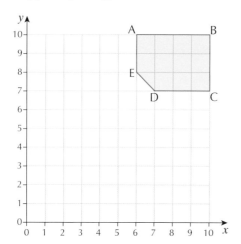

a Rotate the pentagon ABCDE through 90° clockwise about the point (4, 9) to give the image A'B'C'D'E'.

b Write down the coordinates of A', B', C', D' and E'.

c Write down two different rotations that will move pentagon A'B'C'D'E' onto pentagon ABCDE.

6 Copy the triangle onto a coordinate grid with axes for x and y from 0 to 8.

a Rotate the triangle ABC through 90° anticlockwise about the point (4, 4) to give the image A'B'C'.

b Write down the coordinates of A', B' and C'.

c Which coordinate point remains fixed throughout the rotation?

d Fully describe the rotation that will map the triangle A'B'C' onto the triangle ABC.

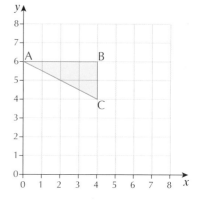

MR **7** Some of these rotations are identical. State which ones are identical and explain why.

90° clockwise

180° clockwise

120° anticlockwise

270° clockwise

240° anticlockwise

450° clockwise

90° anticlockwise

180° anticlockwise

2.4 Translations

1 Describe each of the following translations.

   a   A to C
   b   A to D
   c   C to B
   d   D to E
   e   B to A
   f   B to D
   g   D to C

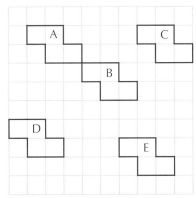

2 Copy the isosceles triangle onto a coordinate grid, with axes for x and y from –5 to 5.

   a   Plot the image A′B′C′ with A′(2, –1), B′(–1, –1) and C′(–1, 2).
   b   Describe the translation to translate ABC to A′B′C′.
   c   Describe the translation to translate A′B′C′ to ABC.

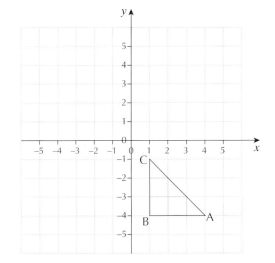

3 To translate shape A to shape B, A is translated 5 units right and 2 units up.

To translate shape B to shape C, B is translated 3 units right and 7 units up.

   a   Describe the translation that translates shape A onto shape C.
   b   Describe the translation that translates shape C back onto shape A.

4 Copy the grid and kites using axes for x and y from 0 to 10.

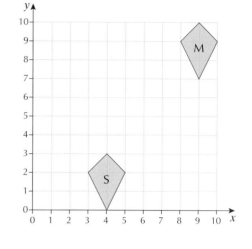

a Write down the coordinates of the vertices of kite M.

b Translate kite M by 2 units left and 6 units down. Label the new kite P.

c Write down the coordinates of the vertices of kite P.

d Translate kite P by 5 units left and 6 units up. Label the new kite Q.

e Write down the coordinates of the vertices of kite Q.

f Translate kite Q by 1 unit right and 3 units down. Label the new kite R.

g Write down the coordinates of the vertices of kite R.

h Describe the translation that maps kite R onto kite S.

5 Copy the isosceles triangle onto a coordinate grid, with axes for x and y from −5 to 5.

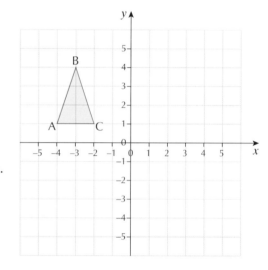

a Translate ABC by 5 units right and 4 units down.

b Write down the coordinates of the image A′B′C′.

c Rotate A′B′C′ by 180° about the point (2, 0). Label the new image A″B″C″.

d Rotate A″B″C″ by 180° about the origin. Label the new image A‴B‴C‴.

e Describe the translation that translates A‴B‴C‴ back onto ABC.

6 Copy the trapezium onto a coordinate grid, with axes for x and y from 0 to 10.

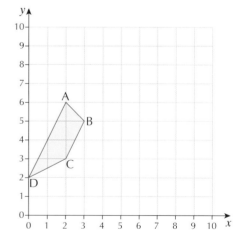

a Translate ABCD by 7 units right and 1 unit up.

b Write down the coordinates of the image A′B′C′D′.

c Rotate A′B′C′D′ by 90° anticlockwise about the point (7, 3). Label the new image A″B″C″D″.

d Describe the transformation that translates A″B″C″D″ back onto ABCD.

Six quadrilaterals have been transformed, either by rotation or translation, from the object ABCD to the image A'B'C'D'.

For each quadrilateral transformation, identify the type of quadrilateral, work out which type of transformation it is and describe the transformation.

Quadrilateral 1: A(2, –1), B(3, –3), C(2, –4), D(1, –3) → A'(–1, 2), B'(–3, 1), C'(–4, 2), D'(–3, 3)

Quadrilateral 2: A(–1, 0), B(–4, 2), C(–1, 4), D(2, 2) → A'(–1, 0), B'(2, –2), C'(0, –4), D'(–4, –2)

Quadrilateral 3: A(3, –3), B(1, 3), C(4, 2), D(5, –1) → A'(–2, –1), B'(–4, 5), C'(–1, 4), D'(0, 1)

Quadrilateral 4: A(4, 2), B(5, –1), C(2, –2), D(1, 1) → A'(–4, 0), B'(–5, 3), C'(–2, 4), D'(–1, 1)

Quadrilateral 5: A(3, 2), B(0, 0), C(1, 2), D(0, 4) → A'(–3, 2), B'(–1, –1), C'(–3, 0), D'(–5, –1)

Quadrilateral 6: A(–2, 3), B(2, 3), C(1, 1), D(–3, 1) → A'(–3, –1), B'(1, –1), C'(0, –3), D'(–4, –3)

For quadrilateral 4, if you were not told which point had which coordinate, what other possible answers would there be for the description of the transformation?

2.5 Constructions

1   **a**   Draw a line AB that is 13 cm long.

Use compasses only to bisect the line.

Check the bisection using your ruler.

b   Draw a line CD that is 9 cm long.

Use compasses only to bisect the line.

Check the bisection using your ruler.

2   **a**   Use a protractor to draw an angle of 64°.

Use compasses to bisect the angle.

Measure the two angles formed with a protractor to check that they are both 32°.

b   Use a protractor to draw an angle of 134°.

Use compasses to bisect the angle.

Check the angles using your protractor.

3   **a**   Draw a line XY that is 12 cm long.
      b   Construct the perpendicular bisector of XY.
      c   By measuring the length of the perpendicular bisector, draw a rhombus with diagonals of length 12 cm and 4 cm.

4 **a** Place your outstretched hand on a sheet of paper. Mark the end of your thumb, middle finger and little finger. Label the points A, B and C respectively.

b Draw the line AC. Bisect it using compasses. Label the midpoint D.

c Compare the width of your closed hand to the length CD.

Step 1   **Step 2**

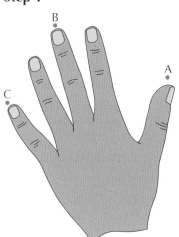 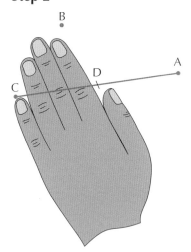

5 **a** Draw a triangle ABC with sides of any length.

b Construct the angle bisectors for each of the three angles.

The three angle bisectors will meet at a point O in the centre of the triangle.

c Using O as the centre, draw a circle to touch the three sides of the triangle.

6 **a** Draw a triangle ABC with sides of any length.

b Construct the perpendicular bisectors for each of the three sides.

The three perpendicular bisectors will meet at a point O in the centre of the triangle.

c Using O as the centre, draw a circle to touch the three corners of the triangle.

(PS)

7 **a** Using only compasses, construct an equilateral triangle.

b Now construct an angle of 15°.

3 Probability

3.1 Probability scales

1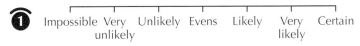

Impossible Very Unlikely Evens Likely Very Certain
unlikely   likely

Copy the scale above. Label the scale with each of the following events.

a You pick one card from a pack of 52 cards and it is an ace (there are four aces in a pack).
b You will experience it raining sometime in the future.
c You are travelling down an unknown road. The next bend is to the left.
d A person can walk on water unaided.
e A wine glass will break when you drop it onto the floor.

2 A child is asked to choose a lucky number from one of the following.

1   2   3   4   5   6   7   8   9

Say which of the following is more likely, or whether they have the same chance.

a even number or number more than 6
b prime number or odd number
c multiple of 5 or multiple of 4
d triangular number or square number

3 Imagine the following quadrilaterals are cut from plastic and placed in a bag.

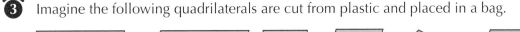

Rectangle   Parallelogram   Square   Rhombus   Kite   Trapezium

You pick one shape out of the bag at random. Copy and complete the following sentences by filling in the missing probability words.

impossible, very unlikely, unlikely, an even chance, likely, very likely, certain

a Picking a shape with parallel sides is ...
b Picking a shape with all equal sides is ...
c Picking a shape with four sides is ...
d Picking a shape with four right angles is ...
e Picking a shape with no equal sides is ...
f Picking a shape with three angles is ...

4 Copy and complete the following table.

Event	Probability of event occurring (p)	Probability of event not occurring ($1 - p$)
A	$\frac{2}{3}$	
B	0.35	
C	8%	
D	0.04	
E	$\frac{5}{8}$	
F	0.375	
G	62.5%	

 5 The probability of an egg having a double yoke is 0.009. What is the probability that an egg does not have a double yoke?

 6 100 rings are placed in a box. Ten are gold, 20 are silver, 36 are plastic and the rest are copper. A ring is chosen at random. What is the probability that it is the following?

    a gold   **b** not silver   **c** copper   **d** not plastic

7 The diagram shows 12 dominoes.

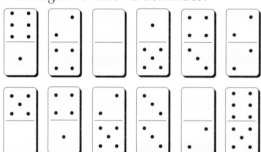

    A domino is chosen at random. Calculate the probability that it:

    a has a 4   **b** does not have a 3   **c** has a total over 5
    d totals less than 10   **e** has a total of 9   **f** is a double whose total
                                                                    is less than 5.

 8 A weather forecaster estimates the probability of fog to be 36%, with a 0.72 probability of rain, but only a 1 in 8 chance of sunshine. What is the probability of the following?

    a no sunshine   **b** not being foggy   **c** no rain

3.2 Mutually exclusive events

 (MR) **1** **a** Draw a Venn Diagram showing the two sets 'square numbers up to 50' and 'numbers smaller than 10'.
      b Which numbers are in both sets?
      c Are the events, 'selecting a square number' and 'selecting a number smaller than 10' mutually exclusive? Explain your answer.

2 In a football league, 8 teams have some red but no blue in their shirts, 4 feature blue but no red, 2 contain black, 3 are all white, the rest have some white. 5 teams have striped shirts.

    a Are red and blue shirts mutually exclusive?

    b Are striped shirts and shirts containing red mutually exclusive?

    c Write down any other pairs of features that you think are mutually exclusive.

3 Andrea picks a card at random from a normal pack of cards. A pack has two red suits and two black suits. Each suit contains 13 cards, ranging from values 1 (ace) to 10 and then jack, queen and king, known as picture cards.

    a Are red cards and picture cards mutually exclusive?

    b Are picture cards and number cards mutually exclusive?

    c Are numbers above 6 and black cards mutually exclusive?

 4 In a game you need to roll a dice and get an odd number less than 5 to win.

    a Draw a Venn Diagram showing the two sets 'odd numbers' and 'numbers less than 5'.

    b Are the events, 'rolling an odd number' and 'rolling a number less than 5' mutually exclusive? Explain your answer.

    c Which number is outside both circles?

    d Which numbers are in the intersection of both circles?

    e Use the Venn Diagram to state the probability of rolling an odd number less than 5.

5 The diagram shows sketches of eight faces.

Which of the following pairs of events are mutually exclusive? (Hint: 'Left eye' means the eye on the left of the diagram.)

    a A smiling face and a sad face.

    b Left eye shut and both eyes shut.

    c Wearing a hat and both eyes open.

    d Both eyes open and a sad face.

    e Wearing a hat and right eye shut.

    f Smiling with an eye open and right eye shut.

6 A six-sided dice is numbered 1, 2, 3, 4, 5 and 6. It is rolled once.

 a From the events below, write down three pairs of events that are mutually exclusive.

 b From the events below, write down three pairs of events that are *not* mutually exclusive.

i number 3	**ii** even number	**iii** number greater than 3
iv square number	**v** triangular number	**vi** multiple of 3
vii prime number	**viii** number 1	

 7 Work out the probabilities for the events in question **6**.

3.3 Sample space diagrams

1 The letters of the word SUCCESSOR are each written on a card and placed in a bag. One of the cards is taken from the bag. What is the probability that the letter is the following?

a s	**b** a vowel	**c** one of the last 10 letters of the alphabet
d c or s	**e** a consonant	**f** a letter of the word ROSE

2 Two children, Kim and Franz, place the following coins in a bag: 1p, 2p, 5p, 10p, 20p, 50p, £1. They then each write their name on a card and put the card in the same bag. A card and a coin are taken from the bag at random. The coin is given to the named person.

 a Make a table to show the possible outcomes.

 b Calculate the probability of the following:

 i Kim receives 20p **ii** Franz receives less than 10p
 iii One of the children receives 10p **iv** Kim does not receive £1.
 v Neither child receives more than 20p

(PS) **3** A taxi firm owns a red and a green taxi cab. The red taxi can carry up to six passengers. The green taxi can carry up to five passengers. Both taxis are in continuous use, that is, they always have at least one passenger.

 a Copy and complete the table showing the total number of passengers being carried at any one time.

		Red taxi			
		1	2	3	...
Green taxi	1	2			
	2				
	3			6	
	...				

 b What is the probability that, at any one time, the number of passengers being carried is the following?

 i 7 **ii** 2 **iii** 12
 iv less than 5 **v** an odd number **vi** 2 or 7

4 An ordinary dice is tossed at the same time as this spinner is spun. Their numbers are added together to give the total score.

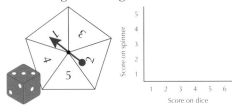

a Make a list of all the possible outcomes. (Axes like the one above may help.)
b How many possible outcomes are there?
c Find each of the following probabilities of scoring:

    i 8   **ii** 1   **iii** an even number   **iv** any score other than 5

    v a score higher than 6
    vi a total where the dice score is lower than the spinner score.

5 Two tetrahedral (four-sided) dice, each numbered 1, 2, 3 and 4, are rolled together. Use a sample space diagram to show all the possible totals when the numbers are added.

(PS) **6** Two girls have three cards each, numbered 1 to 3. One card from each pile is turned over and the numbers are multiplied.
a Design a set of axes like the one in question **4** to find all the possible outcomes.
b What is the probability of getting a total of:

    i 6   **ii** an odd number   **iii** a multiple of 3

    iv a factor of 12   **v** a square number   **vi** a single-digit number?

Brainteaser

Two boys play a game involving a 10-sided dice and a pack of cards. A pack of cards has four suits with 13 cards in each, from ace (1) up to 10, then jack, queen and king.

They agree that an ace counts as 1, jack as 11, queen as 12 and king as 13. They take turns to roll the dice and turn over a new card, alternately.

a If a 5 is rolled, what is the chance that the first card revealed will be:
    i lower   **ii** higher?
b If the 8 of diamonds is turned over, what is the chance that the dice roll is lower?
c What is the probability that the total of the two numbers is:
    i under 6   **ii** over 10   **iii** an odd number?
d One of the boys rolls an even number and, when added to the card, the total value is over 12. What is the probability that the card was:
    i a 5   **ii** a 9?

3.4 Experimental probability

  1 A girl wishes to test her octagonal dice to see if it is biased. She rolls the dice 50 times. The results are shown below.

Score	1	2	3	4	5	6	7	8
Frequency	7	6	7	5	4	3	5	4

    a How often would you expect each number to appear?
    b Do you think the dice is biased? Give a reason for your answer.
    c How could she improve the experiment?
    d From the results, estimate the probability of rolling a 7.
    e From the results, estimate the probability of rolling a 3 or a 4.
    f From the results, estimate the chance of her not rolling a 5.

 2 The numbers of days it rained over different periods are recorded below.

Recording period (days)	Number of days of rain	Experimental probability
30	12	
60	33	
100	42	
200	90	
500	235	

    a Copy and complete the table.
    b What is the best estimate of the probability of it raining? Explain your answer.
    c Estimate the probability of it not raining.
    d Is there a greater chance of it raining or not raining?

 3 Leo's mum thinks they have pizza and baked beans too often at his school.

She asked Leo to keep a record each day for a month of when there was pizza and beans.

At the end of four weeks he gave her these results:

Week 1	Pizza	No pizza
Beans	2	1
No beans	1	1

Week 2	Pizza	No pizza
Beans	1	3
No beans	1	0

Week 3	Pizza	No pizza
Beans	3	0
No beans	1	1

Week 4	Pizza	No pizza
Beans	1	1
No beans	2	1

a Create a summary table showing Leo's results

b **i** Over the month, how many days was pizza served?

　　ii How many days were beans not served?

c What is the probability of the school serving both pizza and beans?

d What is the probability of the school serving neither pizza nor beans?

e What would make this a better sample of the school meals?

  4 A company manufactures items for cars. The number of faulty items is recorded as shown below.

Number of items produced	Number of faulty items	Experimental probability
100	9	
200	22	0.11
500	47	
1000	85	

a Copy and complete the table.

b Which is the best estimate of the probability of an item being faulty? Explain your answer.

  5 Colour 10 identical matchsticks as shown here and place them in a bag.

(If you can't use matchsticks, find some coloured counters, or use small pieces of paper.)

Red　Red　Red　Green　Green　Yellow　Blue　Blue　Blue　Blue

a Take a matchstick from the bag and record its colour in the tally chart below. Do this 10 times.

Colour	Tally	Frequency	Experimental probability	Theoretical probability
Red				
Green				
Yellow				
Blue				

b Calculate the experimental probabilities.
Write your answers as decimals.

c Calculate the theoretical probabilities.
Write your answers as decimals.

d Compare the experimental and theoretical probabilities.

e How could you make sure the experimental and theoretical probabilities are closer?

4 Percentages

4.1 Calculating percentages

  1 Here are some test marks. Write them as percentages.

    a   7 out of 10   **b**   18 out of 25
    c   17 out of 20   **d**   33 out of 50
    e   3 out of 5   **f**   1 out of 20

(FS) **2** Write these amounts of money as percentages.

    a   £19 out of £25   **b**   £13 out of £20
    c   £96 out of £400   **d**   £3 out of £50

 3 Write these quantities as percentages.

    a   3 kg out of 4 kg   **b**   8 cm out of 25 cm
    c   42 km out of 50 km   **d**   102 ml out of 200 ml

 4 Milton drinks 27 cl of a 50 cl bottle of orange.

    a   What percentage did he drink?
    b   What percentage remains?

 5 16 out of 25 pupils in class 8A passed their maths test. 13 out of 20 pupils in class 8B passed the same maths test. Which class had the better results? (Hint: Calculate the percentage for each class.)

 6 Use a calculator to find the following, correct to the nearest per cent.

    a   14 as a percentage of 23   **b**   81 as a percentage of 120
    c   6 as a percentage of 65   **d**   3200 as a percentage of 7000

 7 Delaney drinks 35 cl of an 80 cl bottle of orange.

    a   What percentage did she drink?
    b   What percentage remains?

8 A company employs 280 workers. At the last general election, 130 voted Labour, 83 voted Liberal Democrat and the remainder voted Conservative. Calculate the percentage vote for each party from the company.

9 Ella, Honor and Tia contributed £23, £18.50 and £25.60 toward their total restaurant bill.
   a What percentage of the total bill did each person contribute?
   b If Tia then tipped the waitress 10% of the total bill, what percentage of the new total was contributed by Ella?

10 a Stuart took a maths test lasting 80 minutes. He spent 23 minutes on the Algebra section, 33 minutes on the Number section and the remaining time on the Geometry section. What percentage of the time did he spend on each section?
   b Marcel spent the same amount of time on each section as Stuart, but he also took an Extension section lasting 30 minutes. What percentage of the time did he spend on each section?
   c Sophie was given 25% extra time because she has dyslexia. If she spent the same proportion of her time on each section as Stuart, how long did she spend on each?
   d Compare your answers for parts **a** and **c**. What do you notice? Explain your answer.

4.2 Percentage increase and decrease

1 **a** Increase $40 by 20%.   **b** Decrease 200 kg by 5%.
   c Decrease 70p by 60%.   **d** Increase 2000 m by 25%.
   e Increase 140 ml by 15%.   **f** Decrease £250 by 20%.
   g Increase 230 ml by 45%.   **h** Decrease £7.60 by 85%.

2 Three people each withdrew a certain percentage of their bank balance from their bank account.

John: 29% of £300   Hans: 75% of £640   Will: 15% of £280

Which person has the largest remaining bank balance?

3 'Zipping' a computer file reduces its size by a certain percentage. Find the sizes of these files after zipping.

   a 500kB file reduced by 20%   **b** 740kB file reduced by 5%
   c 840kB file reduced by 25%   **d** 1900kB file reduced by 90%

4 **a** Increase 254 cm^2 by 18%.   **b** Decrease £15.23 by 84%.
   c Increase 7143 g by 47%.   **d** Increase 0.74 litres by 31%.
   e Decrease £17 400 by 97%.   **f** Decrease 715 kg by 28%.

5 Marvin's average score on the computer game *Space Attack* was 256.

    a After buying a new game pad, his average score increased by 28%.
What was his new average score, to the nearest whole number?

    b Marvin went on holiday. When he returned, his average score had decreased by 15%.
What was his new average score, to the nearest whole number?

6 A coat costs £72 before VAT of 17.5% is added.

    a What is the price including VAT?

    b VAT is increased to 20%. How much more expensive is the coat now than when VAT was 17.5%?

7 'Zipping' a computer file reduces its size by a certain percentage.

    a Find the smallest file after zipping.

    b Find which file has been reduced by the most kilobytes.

      500 kB file reduced by 17%

      740 kB file reduced by 43%

      655 kB file reduced by 23%

      1264 kB file reduced by 92%

8 A coat costs £96. It is reduced in a sale by 20%.

    a Find the sale price of the coat.
After the sale, the manager increases the price by 20%.

    b Find the new price of the coat.

    c Explain why the new price of the coat is not the same as the original price of £96.

Brainteaser

The same five candidates stood for the same seat in two local elections.

200 more people voted in the second election than the first.

Victoria received 2800 votes in the first election, which increased by 8% for the second.

Sophie received 4500 votes in the first election, which decreased by 32% in the second election.

Naiga received 6% of the votes in the first election and 23% of the votes in the second election.

Bethany received 1560 votes in the first election, which was 4% more than Naiga.

Honor lost 75% of her votes from the first to the second election.

Who won each election?

Put the candidates in order for each election by the number of votes.

4.3 Percentage change

1 The mass of a cat increases from 8 kg to 10 kg.
   - **a** Show that the multiplier is 1.25.
   - **b** What is the percentage increase?

(FS) **2** The price of a DVD player increases from £250 to £295.
   - **a** Show that the multiplier is 1.18.
   - **b** What is the percentage increase?

3 The number of trees in a forest is increased from 320 to 368.
   Work out the percentage increase.

(FS) **4** Bethany gets a pay rise from £18 500 per year to £19 610 per year.
   Work out the percentage increase.

(FS) **5** Dan bought a calculator from a shop for £23. The shopkeeper paid £20 for the calculator. What was the percentage increase?

(FS) **6** The price of a toaster decreases from $25 to $15.
   - **a** Show that the multiplier is 0.6.
   - **b** What is the percentage decrease?

7 The length of a race decreases from 1500 m to 1260 m.
   - **a** Show that the multiplier is 0.84.
   - **b** What is the percentage decrease?

8 The number of panels on a fence is decreased from 80 to 36.
   Work out the percentage decrease.

(FS) **9** Maria bought a skateboard for £25. She sold it two months later to Zeenat for £18. Zeenat then decorated the skateboard and sold it to Delaney for £27.
   - **a** What was Maria's percentage loss?
   - **b** What was Zeenat's percentage profit?
   - **c** What was the overall percentage change in the value of the skateboard?

(FS) **10** An odd job man kept a record of his income and costs for each job. Copy and complete this table.

Job	Costs (£)	Income (£)	Profit (£)	Percentage profit (%)
4 Down Close	120	162		
High Birches	50			82%
27 Bowden Rd	25		17	
Church hall		34		70%

5 Sequences

5.1 Using flow diagrams to generate sequences

1 Use the flow diagrams below to generate finite sequences.

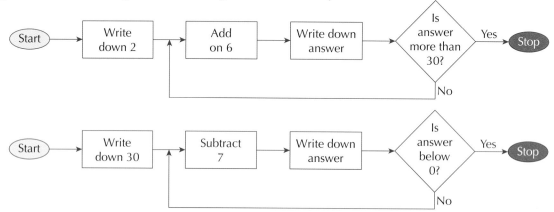

2 a Use the flow diagram below to generate a finite sequence.

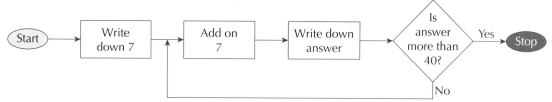

b What do we call this number sequence?

3 a Use the flow diagram below to generate a finite sequence.

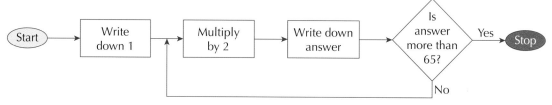

b What do we call this number sequence?

4 a Use the flow diagram below to generate a finite sequence.

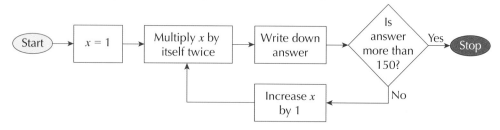

b What do we call this number sequence?

5 Describe how the infinite sequences below are generated.

    a   2, 5, 8, 11, 14, 17, ...   **b**   2, 5, 10, 17, 26, 37, ...

    c   2, 5, 11, 23, 47, 95, ...

 6 The following patterns of dots generate sequences of numbers.

Write down the next four numbers in each sequence.

  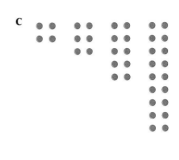

  7 Write down the first five terms for three different sequences beginning
2, 4, ..., ..., ... and explain how each of them is generated.

 8 Describe how each of the following sequences is generated and write down the next
two terms.

    a   20, 21, 23, 26, 30, 35, ..., ...   **b**   50, 49, 47, 44, 40, 35, ..., ...

    c   3, 5, 9, 15, 23, 33, ..., ...   **d**   72, 70, 66, 60, 52, 42, ..., ...

 9 Draw a flow diagram to generate the first six terms for each set of instructions below.

    a   start at 1, add on 4   **b**   start at 1, multiply by 4

    c   start at 5, subtract 2   **d**   start at 360, divide by 2

5.2 Using the nth term of a sequence

 1 For each of the sequences whose nth term is given below, find the following:

    i   the first three terms   **ii**   the 100th term.

    a $3n - 1$ **b** $5n + 2$ **c** $6n - 5$

    d $10n - 1$ **e** $3n + 8$ **f** $\frac{1}{2}n + 1\frac{1}{2}$

 2 The following give the nth term for different sequences.

    a $4n - 1$ **b** $4n + 2$ **c** $4n - 4$ **d** $4n + 5$

For each sequence write down:

    i   the first four terms   **ii**   the first term, a **iii**   the difference, d.

  3 In question **2**, what do you notice about d and the coefficient of n?

4 Here are the nth terms for different sequences.

  a $5n - 1$　　**b** $8n + 2$　　**c** $6n - 9$　　**d** $10n + 3$

  For each sequence write down:

  i the first four terms　　**ii** the difference, d.

 5 For the following sequences, write down the first term, a, and the difference, d.

  a 1, 8, 15, 22, 29,...　　**b** 2, 4, 6, 8, 10,...
  c 3, 12, 21, 30, 39,...　　**d** 6, 3, 0, −3, −6,...

 6 Given the first term, a, and the difference, d, write down the first six terms of each of these sequences.

  a $a = 1, d = 8$　　　　**b** $a = 5, d = 7$　　　　**c** $a = 4, d = -2$
  d $a = 1.5, d = 0.5$　　**e** $a = 10, d = -3$　　**f** $a = 2, d = -0.5$

(PS)   **7** Try to find the nth term for each sequence in question **6**.

5.3 Finding the nth term of a sequence

 1 Each of these sequences uses an 'add' rule. Copy and complete the sequences. Describe the rule used.

  a 2, 6, 10,...,..., 22　　**b** 1, 13,...,..., 49,...　　**c** −10,..., 2, 8,..., 20

2 Work out the general rule (nth term) for the sequences in question **1**. Use that to calculate the 20th term for each.

3 Find the nth term for each of the following patterns. Use this to find the 40th term in each pattern.

  a

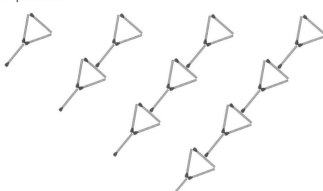

  b

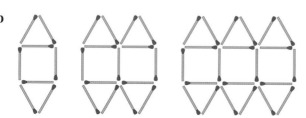

c (Remember that there's a hidden cube in the diagrams below.)

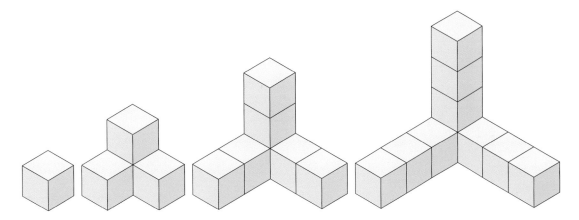

 4 Find the *n*th term of each of the following sequences.

    a   4, 10, 16, 22, 28,...   **b**   8, 11, 14, 17, 20,...   **c**   9, 15, 21, 27, 33,...

    d   4, 7, 10, 13, 16,...   **e**   13, 20, 27, 34, 41,...

 5 In each of the following sequences, find the missing terms, the *n*th term and the 30th term.

Term	1st	2nd	3rd	4th	5th	6th	7th	8th	*n*th	30th
Sequence A	___	___	___	13	16	19	22	___	___	___
Sequence B	___	9	16	___	30	37	___	___	___	___
Sequence C	___	___	25	___	45	___	65	___	___	___
Sequence D	___	11	___	19	___	27	___	___	___	___

 6 In each of the following sequences, find the 5th and 50th terms. (Hint: You will need to work out the *n*th term to do this quickly.)

    a   1, 5, 9, 13,...   **b**   3, 5, 7, 9,...   **c**   4, 12, 20, 28,...

    d   5, 15, 25, 35,...   **e**   2, 8, 14, 20,...   **f**   10, 30, 50, 70,...

    g   2, 5, 8, 11,...   **h**   0, 5, 10, 15,...   **i**   4, 11, 18, 25,...

5.4 The Fibonacci sequence

 1 Fill in the numbers that are missing in the sequence below.

The original Fibonacci sequence

1, 1, 2, 3, 5, —, 13, 21, 34, —, 89, —, 233, 377, ...

 2 What are the next two numbers in the sequence in question **1**?

 3 Which of the first 15 numbers in the original Fibonacci sequence are:

    a   square numbers   **b**   triangular numbers   **c**   prime numbers

    d   multiples of 5   **e**   factors of 48?

4 You can make other Fibonacci sequences by starting with two different numbers.

Write down the first five numbers of a Fibonacci sequence that start with these numbers:

a 1, 2 **b** 1, 4 **c** 4, 4 **d** 3, 7

 5 a Look at the first three consecutive terms from the original Fibonacci sequence:

i Multiply the first and the last numbers together.
ii Multiply the middle number by itself.
iii What is the difference between the numbers?

 b Do the same for the next few sets of three consecutive numbers.

i What do you notice?
ii Can you explain what you notice?

 6 These are excerpts from various Fibonacci sequences. For each sequence work backwards to find:

i the first two non-zero numbers   **ii** the first eight numbers.

a ..., ..., 4, 7, 11 ...   **b** ..., ..., 10, 16 ...   **c** ..., ..., 19, 31 ...

 7 Fibonacci sequences ascend in numerical order. Can any of these pairs of numbers occur in a Fibonacci sequence? If yes, write down the first eight numbers in the sequence:

a ... 15, 20 ...   **b** ... 14, 23 ...   **c** ... 24, 35 ...

Brainteaser

Do the same as question **5** for your sequences in question **4**.

Different things happen each time, but there are patterns to find.

Write down what you notice.

Explore other sequences.

6 Area of 2D and 3D shapes

6.1 Area of a triangle

1 Calculate the areas of the following triangles.

a

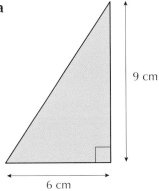

9 cm

6 cm

b

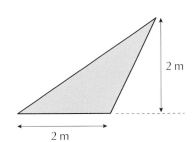

2 m

2 m

c

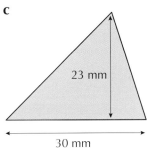

23 mm

30 mm

d

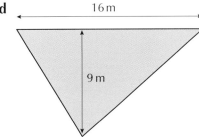

16 m

9 m

e

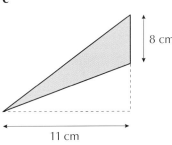

8 cm

11 cm

f

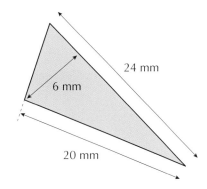

24 mm

6 mm

20 mm

2 Copy and complete the table below which gives the measurements of four triangles.

Base	Height	Area
12 cm	9 cm	
8 cm	14 cm	
6 mm	7 mm	
16 cm		64 cm²
	20 m	100 m²

3 Use squared paper to draw four different triangles with area 24 cm².

(Hint: Draw the base first, using a whole number of centimetres that is a factor of 48, for example, 8 cm. Then calculate the height of the triangle.)

4 Calculate the height of each triangle below using the area given.

a

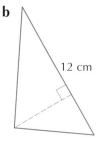

2 m

Area = 10 m²

b

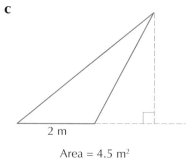

12 cm

Area = 36 cm²

c

2 m

Area = 4.5 m²

5 Each puzzle piece below is a compound shape, and can be put together to make a complete shape.

Each small square represents a square centimetre. How much area will the completed puzzle cover?

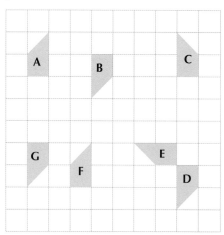

6 Work out the area of the shapes below.

a

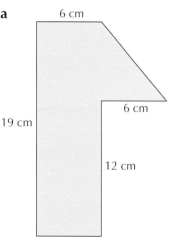

6 cm

19 cm

6 cm

12 cm

b

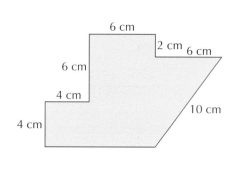

6 cm

2 cm 6 cm

6 cm

4 cm

4 cm

10 cm

7 The shapes below both have an area of 120 cm².

   a In the two rectangles the largest is 8 cm high, so how high is the smaller one?
   b If the base (dotted line) of the triangles measures 5 cm, what is the area of each triangle?
   c What must the height of each triangle be?

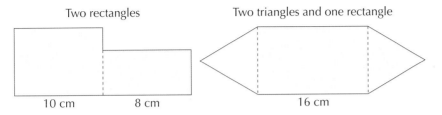

Two rectangles   Two triangles and one rectangle

10 cm   8 cm   16 cm

6.2 Area of a parallelogram

1 Calculate the areas of the following parallelograms.

a

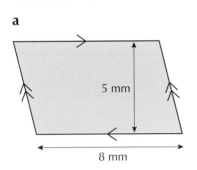

5 mm

8 mm

b

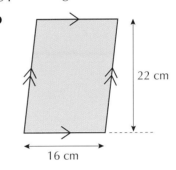

22 cm

16 cm

c

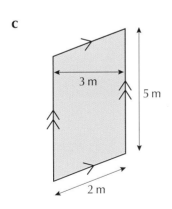

3 m

5 m

2 m

d

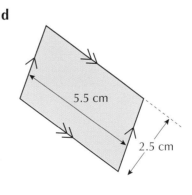

5.5 cm

2.5 cm

2 Copy and complete the table below which gives the measurements of five parallelograms.

Base	Height	Area
7 cm	13 cm	
9 m	19 m	
250 mm	70 mm	
	15 m	120 m²
12 cm		30 cm²

3 Use squared paper to draw four different parallelograms with area 48 cm².

(Hint: Draw the base first, using a whole number of centimetres that is a factor of 48, for example, 8 cm. Then calculate the height of the parallelogram.)

(PS) 4 Each piece is a compound shape or a parallelogram. Each small square represents a square centimetre. Find the area of each puzzle piece. Show your calculations.

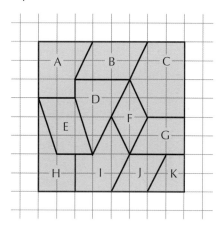

5 Calculate the height of each parallelogram below using the area given.

a

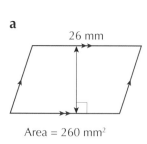

26 mm

Area = 260 mm²

b

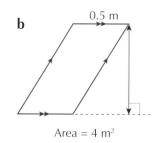

0.5 m

Area = 4 m²

c

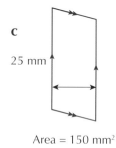

25 mm

Area = 150 mm²

6 Work out **a** the perimeter and **b** the area of this parallelogram.

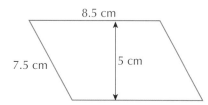

8.5 cm

7.5 cm

5 cm

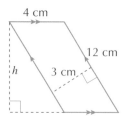

7 Work out the value of *h* in this diagram.

4 cm

12 cm

h   3 cm

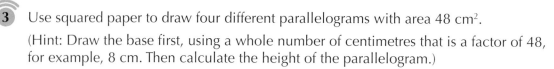

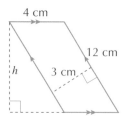

The rhombus shown has an area of 24 cm².

There are five possible whole number values of a and b. Can you find them all?

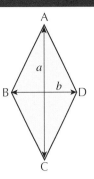

6.3 Area of a trapezium

1 Calculate the areas of the trapezia below.

a

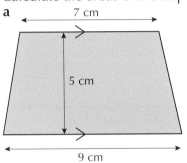

7 cm

5 cm

9 cm

b

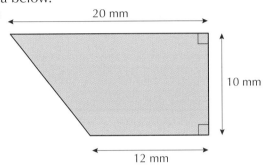

20 mm

10 mm

12 mm

c

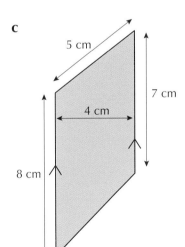

5 cm

7 cm

4 cm

8 cm

d

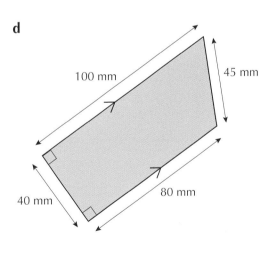

100 mm

45 mm

40 mm

80 mm

2 Copy and complete the table below for trapezia a to e.

Trapezium	Parallel side a	Parallel side b	Height h	Area
a	7 cm	9 cm	3 cm	
b	13 m	8 m	5 m	
c	2 mm	6 mm		32 mm²
d		4 m	6 m	60 m²
e	12 cm		10 cm	250 cm²

3 Use squared paper to draw four different trapezia with area 24 cm².

(Hint: Decide the lengths of the parallel lines first and make sure that their total length is a factor of 48. For example, parallel sides 7 cm and 5 cm have a total length of 12 cm, which is a factor of 48. Then calculate the height of the trapezium.)

 4 Each piece is a compound shape or a trapezium. Each small square represents a square centimetre. Find the area of each puzzle piece. Show your calculations.

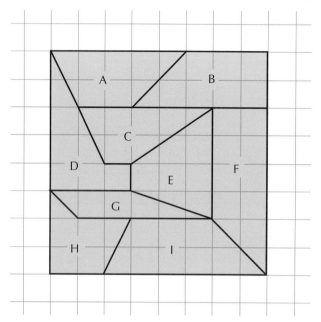

5 Work out the height, h, of the trapezium below if it has an area of:

a 110 cm²   **b** 75cm²

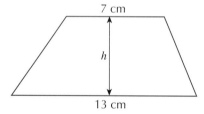

7 cm

h

13 cm

6 The lean-to shed shown below has a cross-sectional area of 4.5 m². How high is it at its tallest point?

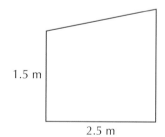

1.5 m

2.5 m

7 The area of this trapezium is 9 cm². Work out three different whole number values of a, b and h, with $b > a$.

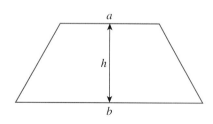

a

h

b

6.4 Surface area of cubes and cuboids

1 Find the surface area of each of the following cuboids.

a

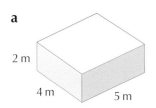

2 m, 4 m, 5 m

b

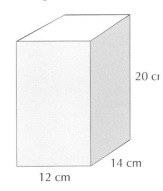

20 cm, 14 cm, 12 cm

c

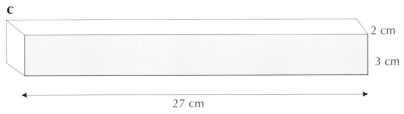

2 cm, 3 cm, 27 cm

2 Calculate the surface area of a cube with sides 6 cm.

3 If a cube has a surface area of 150 cm², what is the length of each side?

4 Krispies are sold in three sizes: mini, medium and giant.

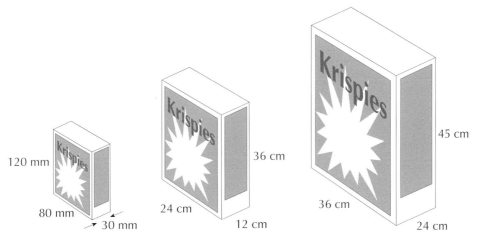

120 mm, 80 mm, 30 mm

36 cm, 24 cm, 12 cm

45 cm, 36 cm, 24 cm

a Calculate the surface area of each box.
(Hint: Convert millimetres to centimetres.)

b How many times more cardboard is needed to manufacture the medium box compared to the mini box (to the nearest whole number)? And how many times for the giant compared to the medium, to one decimal place?

c The area of the logo on the packet front of each box is as shown below:

mini box – 40 cm²   medium box – 448 cm²   giant box – 900 cm²

Work out the fraction taken up by the logo each time in its simplest form.

5 Calculate the surface area of each of the following 3D shapes.

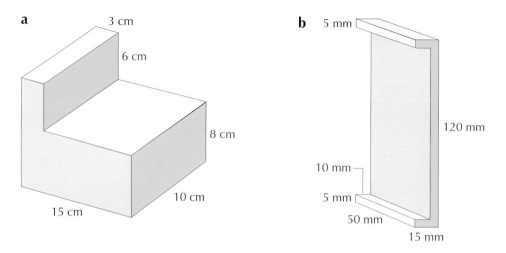

a
3 cm
6 cm
8 cm
10 cm
15 cm

b
5 mm
120 mm
10 mm
5 mm
50 mm
15 mm

Brainteaser

A scout group constructs a wooden log store on a concrete base to keep their wood supply dry. The storage is a cuboid shape, 2.5 m long, 2 m wide and 1 m high with a hinged lid. To keep it waterproof it has to be coated in wood stain.

a What area of wood has to be covered with   **i** 1 coat?   **ii** 2 coats?

Water-based wood stain can be bought at £24.99 for a 2.5 litre tin. The tin claims to cover 8 m² per litre and recommends applying two coats.
Oil-based tins cost £45.99 for 2.5 litres, can cover 20 m² but only needs one coat.

b How much area can the water-based tin cover in total? How many tins are needed?
c How much will be left over from the water-based tins after two coats are applied?
d How many square metres does the oil-based tin cover per litre?
e Which tin works out cheapest and by how much per litre?

7 Graphs

7.1 Graphs from linear equations

1 a Copy and complete this table for the function $y = x + 5$.

x	0	1	2	3	4	5
$y = x + 5$	5	6	7	8	9	10

   b Draw a grid with x-axis from 0 to 5 and y-axis from 0 to 10.
   c Draw the graph of the function $y = x + 5$.

2 a Copy and complete this table for the function $y = x - 3$.

x	0	1	2	3	4	5	6
$y = x - 3$	−3	−2	−1	0	1	2	3

   b Draw a grid with x-axis from 0 to 6 and y-axis from −5 to 5.
   c Draw the graph of the function $y = x - 3$.

3 a Copy and complete this table for each of the functions.

x	0	1	2	3	4	5
$y = x$	0	1	2	3	4	5
$y = 2x$	2	3	4	5	6	7
$y = 3x$	3	4	5	6	7	8
$y = 4x$	4	5	6	7	8	9

   b Draw a grid with x-axis from 0 to 5 and y-axis from 0 to 20.
   c Draw the graph of each function in the table using the same grid.
   d What is different about the lines?
   e Use a dotted line to sketch the graph of $y = 2.5x$.

4 a Copy and complete this table for each of the functions.

x	−2	−1	0	1	2	3
$y = x + 2$						
$y = 2x + 2$						
$y = 3x + 2$						
$y = 4x + 2$						

   b Draw a grid with x-axis from −2 to 3 and y-axis from −10 to 15.
   c Draw the graph of each function in the table.
   d What is the same about the lines?
   e What is different about the lines?
   f Use a dotted line to sketch the graph of $y = 2.5x + 2$.

5 **a** Copy and complete this table for each of the functions.

x	−2	−1	0	1	2	3
$y = 5x - 1$						
$y = 2x - 4$						

b Write down the coordinates of the point where the lines intersect.

6 **a** Draw a grid with x-axis from −2 to 3 and y-axis from −15 to 15.
b Draw the graph of the function $y = 3x + 1$.

7 For the function $8y + 6x = 24$, do the following.
a Find y when $x = 0$.
b Find x when $y = 0$.
c Find a third point that lies on the line.
d Draw the graph of the function.

8 Draw the graph of the function $10y + 4x = 20$ by finding three points that lie on the line.

7.2 Gradient of a straight line

1 For each of these lines, write down the following.
 i the gradient   **ii** where it cuts the y-axis

a 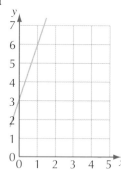 **b** 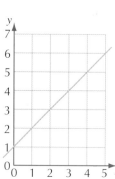 **c** 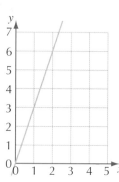 **d**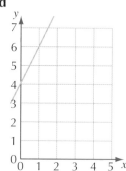

2 Write down the equation of each line in question **1**.

3 State the gradient of each straight line.
 a $y = 7x + 2$ **b** $y = 2x - 8$ **c** $y = x + 5$

4 State the equation of the straight line with:
 a a gradient of 5, passing through the y-axis at (0, 8)
 b a gradient of 6, passing through the y-axis at (0, 11)
 c a gradient of 4, passing through the y-axis at (0, 5).

5 Think about the line joining (0, 3) and (3, 9).

    a What are the coordinates of the point where this line cuts the y-axis?
    b By plotting both points on a coordinate grid, calculate the gradient of this line.
    c What is the equation of this line?

6 Think about the line joining (2, 1) and (4, 7).

    a By plotting both points on a coordinate grid, calculate the gradient of this line.
    b What is the coordinate of where this line cuts the y-axis?
    c What is the equation of this line?

7 What are the equations of the lines that pass through the coordinates:

    a (2, 8) and (4, 10)   **b** (2, 1) and (5, 7)
    c (1, 5) and (2, 9)   **d** (1, 1) and (3, 7)

8   **a** Plot the points A(1, 3) and B(3, 5).
    b Work out the gradient of AB.
    c Extend the line to cross the y-axis and give this point.
    d Write down the equation of the line that passes through AB.

9 Which of these points lie on the line with the equation $y = 2x + 3$?

    a (1, 4)   **b** (2, 7)   **c** (3, 9)
    d (4, 12)   **e** (5, 13)   **f** (6, 17)

Brainteaser

Five lines have been drawn on a piece of paper.

All five lines have positive gradients.

   Line A passes through the points (1, 2) and (3, 8).

   Line B is parallel to line E.

   Line C intersects line D at the point (2, 9).

   Line E has the equation $y = 2x + 7$.

   Line B intersects line C at the point (1, 3).

   Line C is twice as steep as line A.

   Line B intersects the y-axis at the same point as line D.

Find the equation of line D.

7.3 Graphs from simple quadratic equations

 1 Copy and complete the table for $y = x^2 + 5$.

x	−3	−2	−1	0	1	2	3
x^2							
$y = x^2 + 5$							

 2 Copy and complete the table for the equations shown.

x	−3	−2	−1	0	1	2	3
$y = x^2 + 6$							
$y = x^2 + 7$							
$y = x^2 + 8$							

(MR) **3**
- **a** Draw a grid with x-axis from −3 to 3 and y-axis from −1 to 20.
- **b** Draw the graph of each function in question **2**.
(PS)
- **c** What is the same about the lines?
- **d** What is different about the lines?
- **e** Use a dotted line to sketch the graph of $y = x^2 + 9$.

 4
- **a** Copy and complete the table for the equations shown.

x	−3	−2	−1	0	1	2	3
$y = x^2$							
$y = 2x^2$							
$y = 3x^2$							

- **b** Draw a grid with x-axis from −3 to 3 and y-axis from −1 to 30.
- **c** Draw the graph of each function in the table.
- **d** Use a dotted line to sketch the graph of $y = 1.5x^2$.

 5 Copy and complete the table for $y = 2x^2 + 1$.

x	−3	−2	−1	0	1	2	3
x^2							
$2x^2$							
$y = 2x^2 + 1$							

(MR) **6**
- **a** Copy and complete the table for $y = x^2 − 1$.

x	−3	−2	−1	0	1	2	3
x^2							
$y = x^2 − 1$							

- **b** Explain why there are two values of x for which $x^2 − 1$ is equal to 3.

(MR) **7** Rosie and Emma are discussing the equation $y = x^2 + 4$.

Rosie says that when $x = −2$, $y = 0$.

Emma says that when $x = −2$, $y = 8$.

Who is correct and why?

8 A ball is dropped from a tall building. The distance (D metres) travelled by the ball is related to the time (T seconds) it has been falling by the equation $D = 5T^2$.

    a Draw a graph to show how far the ball has fallen for times from 0 to 4 seconds.

    b The building is 100 m tall. Find the time it takes the ball to fall until it is 40 m above the ground.

  9 Use the tables to find the equations of the lines:

x	−3	−2	−1	0	1	2	3
Line 1: $y =$	19	14	11	10	11	14	19
Line 2: $y =$	7	2	−1	−2	−1	2	7
Line 3: $y =$	81	36	9	0	9	36	81
Line 4: $y =$	−1	−6	−9	−10	−9	−6	−1
Line 5: $y =$	31	16	7	4	7	16	31

7.4 Real-life graphs

 1 Mrs Jay had to travel to a job interview at Penford, which was 120 miles away. She caught the 09:00 train from Shobton and travelled to Deely, 30 miles away. This train journey took 1 hour.

There she had to wait 30 minutes for a connecting train to Penford. This train took 1 hour.

Her job interview at Penford lasted 2 hours.

Her return journey to Shobton lasted $1\frac{1}{2}$ hours.

    a Copy this grid, using a scale of 2 centimetres to 1 hour and 1 centimetre to 10 miles.

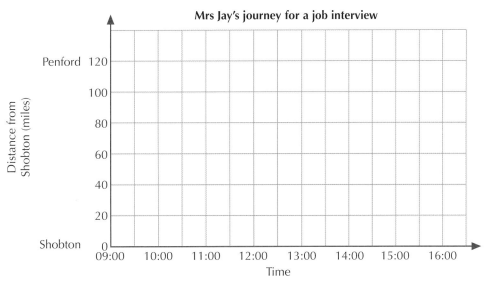

    b Draw on the grid a distance–time graph for Mrs Jay's journey.

    c Mark Deely on the vertical axis.

    d How far was Mrs Jay from Shobton at 1130?

    e At which times was Mrs Jay 60 miles from Shobton?

2 A petrol tanker made this journey along a motorway.

Filled up at petrol depot.

Drove 10 miles in 30 minutes to Dibley Service Station.

Spent 30 minutes filling the pumps.

Drove a further 20 miles in 1 hour to Penton Service Station.

Spent 30 minutes filling the pumps and 30 minutes for a tea break.

Drove a further 30 miles in 1 hour to Hillview Service Station.

Spent 30 minutes filling the pumps.

Returned to the depot in 90 minutes.

a Draw a grid with the following scales.

horizontal axis (time) from 0 to 6 hours, 1 centimetre to 30 minutes

vertical axis (distance) from 0 to 60 miles, 1 centimetre to 5 miles

b Draw on the grid a distance–time graph for the journey.

c Mark the places of delivery on the vertical axis.

d How far was the tanker from the depot after:

i 90 minutes   **ii** 3.5 hours   **iii** 5 hours?

3 **a** Draw a grid with the following scales.

horizontal axis representing time from 0 to 6 hours, 1 centimetre to 30 minutes

vertical axis representing distance from factory from 0 to 50 miles, 1 centimetre to 5 miles

b Draw on the grid a distance–time graph for the following journey. Mark the places of delivery on the vertical axis.

A car transporter left the factory and took 30 minutes to travel 15 miles to a car dealer in Harton. It took half an hour to unload three of the cars. The transporter travelled at 25 mph for another hour and made a delivery at Glimp. This delivery and lunch took an hour. A final delivery was made 30 minutes later at Unwich after a 10-mile drive. This delivery took 30 minutes. The transporter returned to the factory, taking two hours to get back.

c Calculate the average speed of the transporter:

i between Glimp and Unwich   **ii** on the return journey.

4 **a** Draw a grid with the following scale.

time on the horizontal axis showing 0 to 30 minutes, 1 centimetre to 5 minutes

depth on the vertical axis showing 0 to 150 centimetres, 2 centimetres to 50 centimetres

b A swimming pool, 1.5 m deep, was filled with water from a hose. The pool was empty at the start and the depth of water in the pool increased at the rate of 5 centimetres/minute. Copy and complete the following table, showing the depth of water at various times.

Time (minutes)	0	5	10	15	20	25	30
Depth (centimetres)							

c Draw a graph to show the increase in depth of water against time.
d The next day the pool was filled with a different hose that poured in water at a rate of 6 centimetres/minute. Draw a graph to show this.
e Explain how your graph shows which hose filled the pool more quickly.

Brainteaser

Miss Stone forgot to buy some doughnuts for her meeting at 10:45.

The shop was 6 kilometres from her office.

She set off driving to the shop at 10:15.

For the first 4 minutes, the distance travelled after T minutes was given by $D = \frac{1}{8}T^2$.

She then travelled at a steady speed of 30 km/h until she arrived at the shop.

She spent 5 minutes in the shop.

On the return journey she drove the first 2 kilometres at an average speed of 40 km/h.

She was stuck at traffic lights for 2 minutes and then returned to the office at a steady speed of 20 km/h.

How many minutes late was she for the meeting?

Simplifying numbers

8.1 Powers of 10

1 Multiply each of these numbers by   **i** 10   **ii** 100   **iii** 1000.

  a 2.7   **b** 0.05   **c** 38   **d** 0.008

2 Multiply each of these numbers by   **i** 10^4 **ii** 10^5.

  a 2.7   **b** 0.05   **c** 38   **d** 0.008

3 Divide each of these numbers by   **i** 10^4 **ii** 10^5.

  a 730   **b** 4   **c** 2.8   **d** 35 842

4 Work your way along this chain of calculations for each of these starting numbers.

  a 2000   **b** 7   **c** 0.06

  i

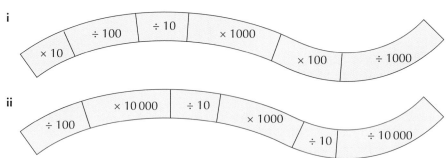

  ii

5 The mass of one proton is 0.000 000 000 000 000 000 000 001 7 grams.

  a What is the mass of one thousand protons?
  b What is the mass of one hundred million protons?

6 The mass of the Earth is 6 000 000 000 000 000 000 000 000 kg.

  a Write the mass of the Earth in grams.
  b The moon is about 1.2% of the mass of the Earth. Write the mass of the moon in kilograms.

7 Light travels at 300 000 000 metres per second.

  a How many metres does light travel in 10^7 seconds?
  b How many kilometres does light travel in one second?
  c How many millimetres does light travel in 10 seconds?

(PS) **8** A rocket travels at 40 000 miles per hour.

  How far will the rocket travel in

  a 10 000 hours   **b** 10^3 hours   **c** 100 weeks

8.2 Large numbers and rounding

 1 Round these numbers to the nearest hundred.

   a 2620   **b** 861   **c** 83 717   **d** 4934

 2 This graph shows the numbers of oil shares sold every hour during a trading day.

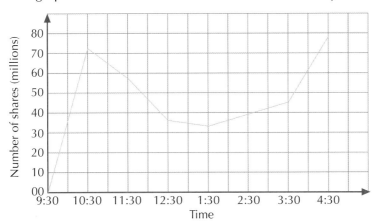

Estimate the number sold each hour. Make a table for your answers.

 3 Round these numbers:

   i to the nearest ten thousand

   ii to the nearest hundred thousand

   iii to the nearest million.

   a 7 247 964   **b** 1 952 599   **c** 645 491   **d** 9 595 902

4 The top six UK airports, in 2006, measured by number of passengers are shown in the table below.

1	Heathrow	67 339 000
2	Gatwick	34 080 000
3	Stansted	23 680 000
4	Manchester	22 124 000
5	Luton	9 415 000
6	Birmingham	9 056 000

   a Which two of these airports had the same number of passengers to the nearest million?

   b How many passengers used Heathrow, Gatwick or Stansted in 2006?
   Give your answer to the nearest ten million passengers.

 5 The populations of four towns in Yorkshire are recorded in the table.

What are the highest and lowest figures that the population could be for each town?

1	Leeds	761 100
2	Sheffield	552 000
3	Scunthorpe	72 700
4	Grimsby	88 000

  6 A chairperson said: "We have about 25 000 members in our society."

An opponent of the society said that they only had 20 000 members.

State a number of members for which both the chairperson and the opponent are correct and explain how each person rounded it.

 7 The mass of the Sun is about 2 000 000 000 000 000 000 000 000 000 000 kg.

a Using this information, what are the highest and lowest masses it could be?

b A more accurate value for the mass of the Sun is about 1 990 000 000 000 000 000 000 000 000 000 kg. What are the highest and lowest masses it could be for this value?

Brainteaser

The table shows the populations of five countries in 2000 and in 2010.

The populations are all rounded to the nearest million.

Country	Population in 2000	Population in 2010
United Kingdom	58 000 000	62 000 000
Netherlands	16 000 000	17 000 000
France	59 000 000	63 000 000
Spain	43 000 000	46 000 000
Italy	58 000 000	60 000 000

c What are the highest and lowest figures that the population of Italy could be for each year?

d What is the maximum possible percentage increase for the population of Spain between 2000 and 2010??

e What is the minimum possible percentage increase for the population of France between 2000 and 2010??

f Find which countries have the minimum and maximum possible percentage increases.

8.3 Significant figures

 1 State the number of significant figures in each of the following.

    a 40   **b** 35 200   **c** 900 000   **d** 470 000

2 Round these numbers to one significant figure.

    a 410   **b** 430   **c** 450   **d** 470

3 State the number of significant figures in each of the following.

    a 4.5478   **b** 6 302 000   **c** 0.04923   **d** 5 billion

4 Round these numbers to one significant figure.

a 8.265   **b** 6.849   **c** 3.965

d 0.095   **e** 4.994   **f** 0.047

 5 Mount Everest is 8848 m tall.

Round the height of Mount Everest to:

a one significant figure

b two significant figures

c three significant figures.

 6 Use a calculator to work out the following divisions. Then write down each answer to three significant figures.

a 1 ÷ 19   **b** 2 ÷ 19   **c** 3 ÷ 19   **d** 4 ÷ 19

e 5 ÷ 19   **f** 6 ÷ 19   **g** 7 ÷ 19   **h** 8 ÷ 19

 7 A football team scored 98 goals in 38 matches. Find, to two significant figures, the mean number of goals scored.

 FS **8** Find the annual salary, to four significant figures, of the following people who have a monthly salary of:

| Rebecca | £2673.48 | Tia | £4711.50 |
| Zeenat | £3188.13 | Tiara | £4273.76 |

 PS **9** A wall measures 19.52 m by 10.45 m. Rosie has four cans of paint and each can has enough paint to coat 45 m². Use estimation to determine if she has enough paint.

8.4 Standard form with large numbers

 1 Write each number as a power of 10.

a 100   **b** 1 000 000   **c** 10 000

d 10   **e** 1 000 000 000 000 000   **f** 100 000

 2 State for each number whether it is written in standard form and if not, explain why.

a 6.8   **b** 0.68×10^{12} **c** 6.8×10^{12}

d 6.8×9^{12} **e** 68×10^{12} **f** 6.8×10^{12}

 3 Write each of the following numbers in standard form.

a 47 800   **b** 720 000   **c** 29 600 000

d 9428   **e** 84   **f** 11 000 000 000

 4 The populations of six South American countries in 2013 are given below. Write each of the numbers in standard form.

a Brazil: 201 000 000   **b** Colombia: 47 100 000   **c** Venezuela: 29 760 000

d Paraguay: 6 800 000   **e** Uruguay: 3 300 000   **f** Guyana: 798 000

5 Write each of the following numbers in standard form.

    a 7.1 million   **b** 467 million   **c** 32 thousand

    d six hundred thousand   **e** 9.9 thousand   **f** quarter of a million

 6 In England, a billion used to be a million million. Write each of the following in standard form using the old billion.

    a 4 billion   **b** 35 billion   **c** 873.2 billion

    d half a billion   **e** a million billion   **f** 62 billion billion

 7 Each of the measurements below relating to the Caspian Sea is written in standard form. Write them as ordinary numbers.

    a Maximum depth = 1.025×10^3 m

    b Length = 1.2×10^6 m

    c Area = 3.71×10^{11} m^2

    d Water volume = 7.82×10^{13} m^3

 8 $2^3 = 2 \times 2 \times 2 = 8$. Find the cube of each number below, giving your answer in standard form.

    a 40

    b 2 million

    c 35 thousand

 9 Write each of the following numbers in standard form.

    a 52×10^7 **b** 342×10^4 **c** 0.6×10^{10}

    d 0.007×10^{20} **e** 3600×10^3 **f** 28×10^{74}

 10 At the end of 2013, the GDP of the United States was 1.6×10^7 and the GDP of the United Kingdom was 2.4×10^6. Approximately how many times larger was the USA GDP than the UK GDP?

Brainteaser

Put these shapes in order of size according to their areas or surface areas, from smallest to largest.

Area

  Rectangle: Base of 5×10^4 cm and height of 1.3×10^2 cm

  Square: Sides of 7.9×10^3 cm

  Triangle: Base of 2.8×10^3 cm and height of 4.7×10^4 cm

  Parallelogram: Base of 4.8×10^5 cm and height of 15 cm

Surface area

  Cuboid: Edges of 2.3×10^3 cm, 4.1×10^3 cm and 3.6×10^3 cm

  Cube: Edges of 3.3×10^3 cm

8.5 Multiplying with numbers in standard form

 1 Work out each of the following and give your answers in standard form.

   a 5×8 **b** 200×3 **c** 4000×7

   d 6×3 **e** 39×10 **f** 5^2

 2 Work out each of the following and give your answers in standard form.

   a 500×900 **b** $70\,000 \times 800$ **c** 400^2

   d 6000×1500 **e** $3\,000\,000^3$ **f** $700\,000 \times 900\,000\,000$

 3 Work out each of the following and give your answers in standard form.

   a $(3 \times 10^5) \times (2 \times 10^3)$ **b** $(2 \times 10^4) \times (2 \times 10^6)$ **c** $(6 \times 10^9) \times (4 \times 10^3)$

   d $(9 \times 10^9) \times (7 \times 10^7)$ **e** $(4 \times 10^7)^2$ **f** $(9 \times 10^{12})^2$

 4 Work out each of the following and give your answers in standard form. Do not round your answers.

   a $(3.7 \times 10^3) \times (6.1 \times 10^2)$ **b** $(5.9 \times 10^6)^2$

   c $(4.8 \times 10^5) \times (6.2 \times 10^4)$ **d** $(6.88 \times 10^7) \times (2.9 \times 10^5)$

 5 Work out each of the following and give your answers in standard form. Round your answers to three significant figures.

   a $(5.28 \times 10^6) \times (4.73 \times 10^4)$ **b** $(4.85 \times 10^3)^2$

   c $(7.72 \times 10^8) \times (2.88 \times 10^4)$ **d** $(9.15 \times 10^2) \times (3.44 \times 10^7)$

   e $(3.91 \times 10^5)^2$

  6 In this question, a, b, c and d should all be positive whole numbers below 10. Write down possible values for a, b, c and d such that:

   a $(a \times 10^b) \times (c \times 10^d) = 6 \times 10^7$

   b $(a \times 10^b) \times (c \times 10^d) = 2.8 \times 10^9$

   c $(a \times 10^b)^2 = c \times 10^d$

  7 Find the total of $50 \times (3.1 \times 10^4) + 200 \times (4.2 \times 10^3) + 4000 \times (1.2 \times 10^2)$. Write the answer in standard form.

   8 Work out each of the following and give your answers in standard form.

   a $(3 \times 10^6)^3$ **b** $(2 \times 10^9)^5$

   c the square root of 1.44×10^{10} **d** the square root of 2.5×10^{37}

 9 Work out each of the following and give your answers in standard form.

   a $(8 \times 10^8) \div (2 \times 10^2)$ **b** $(6.2 \times 10^{11}) \div (3.1 \times 10^4)$ **c** $(3 \times 10^{10}) \div (6 \times 10^5)$

9 Interpreting data

9.1 Pie charts

1 150 children went on one of four school summer holidays. How many children chose the following holidays?

   a camping   **b** France

   c pony trek   **d** Disney World

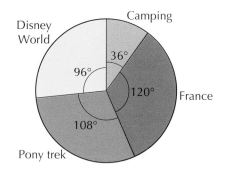

2 One weekend on a train, the staff sold 1200 drinks.

The pie chart illustrates the different drinks that were sold.

How many of the following drinks were sold that weekend?

   a water   **b** tea

   c hot drinks   **d** cold drinks

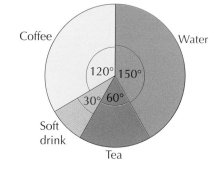

3 Mona carried out a survey on how often people ate chocolate. She asked 72 people.

The pie chart illustrates her results.

Measure the angles to work out how many of these people did the following:

   a never ate chocolate

   b ate chocolate once a week

   c ate chocolate occasionally

   d ate chocolate every day

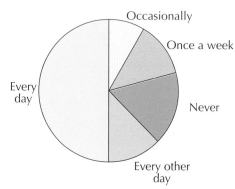

4 Motorbike registrations in October 2013 fell into the categories shown in the pie chart.

To the nearest hundred, there were 6000 bikes registered in total.

   a You have to be at least 17 to ride a bike bigger than 50 cc. How many bikes were registered to people aged 17 or over?

   b 16-year-olds can only legally ride a bike or moped up to 50 cc. What proportion of bikes were registered to 16 year olds that month?

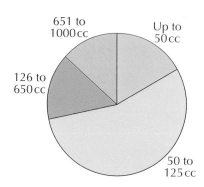

5 A dog rescue centre provides refuge for four types of dog as shown in the chart below.

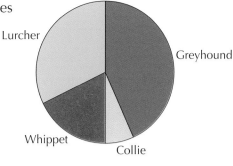

They currently have six collies.

  a Estimate the angles for each sector and work out how many of each breed there are and how many dogs they have in total.

  b Now measure the angles to see how close you were (don't cheat yourself by measuring first!).

6 An international event was attended by people from five European countries as shown in the diagram.

  a Express each country's attendance as a fraction in its simplest terms.

  b If there were 240 Germans, how many people attended altogether?

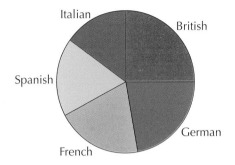

7 These pie charts show how two companies first contacted new customers.

  a Work out how many letters each company sent.

  b Which company sent the most emails? Explain your answer.

  c Make two more comparisons between the companies.

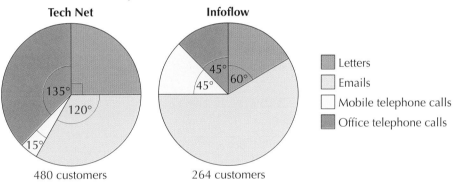

Tech Net   **Infoflow**

480 customers   264 customers

☐ Letters
☐ Emails
☐ Mobile telephone calls
☐ Office telephone calls

Brainteaser

This pie chart shows an approximation of the size of the eight major planets in our solar system in relation to each other.

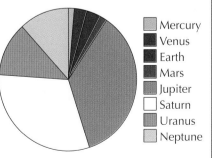

☐ Mercury
☐ Venus
☐ Earth
☐ Mars
☐ Jupiter
☐ Saturn
☐ Uranus
☐ Neptune

a Would this data be better as a bar chart? Can you convert the pie chart to a suitable bar chart?

b Roughly how many times bigger than Mars is Neptune?

c Find two planets where one is about four times larger than the other.

d An approximate way to find the actual diameter of each planet is to multiply the angle by a thousand. The angle for Mercury is 5°, so its diameter is about 5000 km. However, there is an error of up to 10% for each planet. Use this information to find:

  i the smallest possible diameter of Venus   **ii** the largest possible diameter of Saturn.

9.2 Creating pie charts

1 The results of a transport survey show the various ways students go school.

Copy the table and fill in the gaps. Then draw a pie chart to show the results.

Transport	Frequency	As 120 pupils are represented by 360°, each pupil will be represented by 360 ÷ 120 = °
Car	23	23 × 3 = 69°
Bus	17	17 × 3 =
Train	25	25 × =
Bicycle		× 3 =
Walk	40	× =
Total	120	Check the total = 360°

2 Draw a pie chart to represent the numbers of birds spotted on a field trip.

Bird	Crow	Thrush	Starling	Magpie	Other
Frequency	19	12	8	2	19

3 **a** Draw a pie chart to represent the sizes of dresses sold in a shop during one week.

Size	8	10	12	14	16	18
Frequency	3	7	10	12	6	2

    b What size are a quarter of the dresses?
    c What fraction of dresses are size 16 or above?
    d What percentage are size 14?

  4 Trains arriving at Blackpool station were monitored to see how close they ran to time. The results are shown in the table.

     a Draw a pie chart to display them.

Early	4
On time	18
Up to 5 min late	14
5 to 10 min late	3
Over 10 min late	1

    b How many trains were late?
    c What proportion were early?
    d The railway promises that no more than 10% of trains will be over 5 min late, and no more than 5% will be over 10 min late. Write a brief report to explain whether or not they achieved this.

5 A player entered a golf tournament with 18 holes and the table shows how well she did on the first day: *(par is the target, birdie is 1 less, eagle 2 less, bogey is 1 over par, double bogey is 2 over par)*

Eagle	1
Birdie	4
Par	11
Bogey	2
Double bogey	0

a Draw a pie chart to illustrate her results.
b Was her overall score better than par or worse? By how much?
c Par for one round on this course is 72. Major tournaments play four rounds of 18 holes. If she repeats these results each day, what would her total score be?

6 Match the pie charts with their equivalent bar charts:

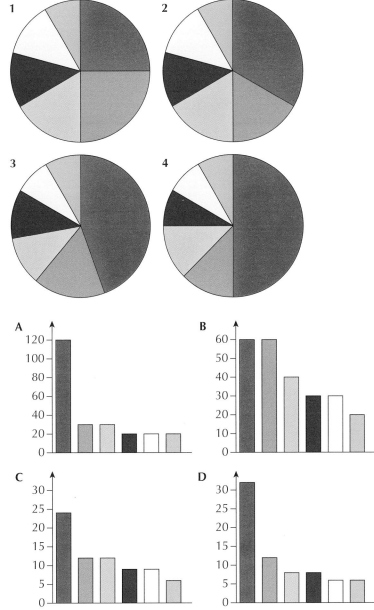

9.3 Scatter graphs and correlation

1 Decide which of these correlations is best applied to each diagram below:

strong positive   strong negative   moderate positive

moderate negative   weak positive   weak negative   none

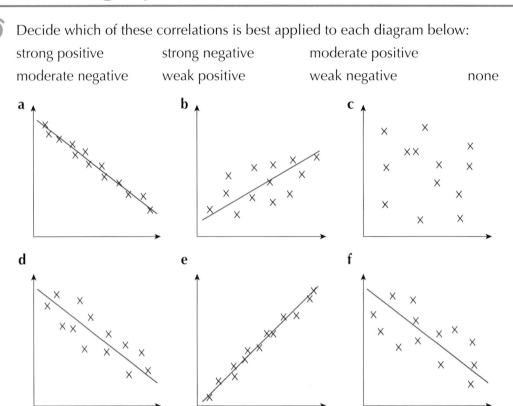

2 A study into various books was carried out, looking at the number of pages, number of chapters and price. The following scatter diagrams were created from the results.

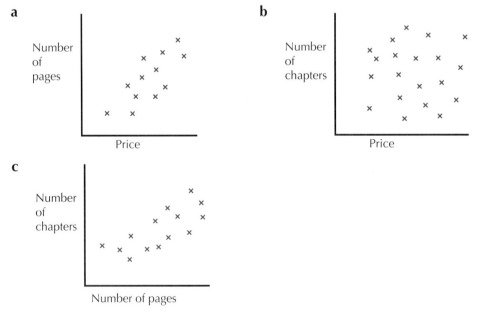

Describe the type of correlation and what each graph tells you.

3 After a study into people's ages, weight, height and the average hours of sleep they get, these scatter diagrams were created.

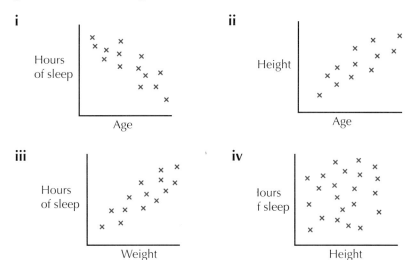

i

Hours of sleep / Age

ii

Height / Age

iii

Hours of sleep / Weight

iv

Hours of sleep / Height

a Match each graph to the statements below:
1 The more you weigh, the more sleep you need.
2 Your height does not affect how much sleep you need.
3 The older you are, the taller you are.
4 The older you are, the less sleep you need.

b One of these graphs would level off if the x-axis was extended further. Which do you think it is? Why?

4 The scatter graph shows scores for students who did Maths and Science tests in the same week.

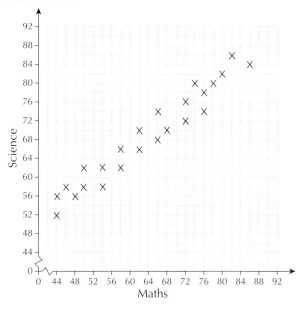

a How many students did both tests?
b From these results how much would you expect a student to score:
  i in Science if they got 82 in Maths   **ii** in Maths if they got 64 in Science?
c Is it fair to say students who are good at Maths are also good at Science?
d Is this group of pupils generally better at Maths or Science? Explain your answer.

5 Which of these will go in each category in the table? Which don't belong in any category?

    a The hotter the weather, the more ice creams are sold.
    b The further you drive, the less fuel you have left in your tank.
    c The more sweet food you eat, the more visits to the dentist you may need.
    d The higher the price of perfume, the more it will rain.
    e The taller you are, the less time it takes to run 400 m.
    f The more revision you do, the better your exam results will be.

	Strong	Weak
Positive correlation		
Negative correlation		

6 The table records how many comedy programmes were shown on BBC channels each evening for a week and the number of times a mountain rescue team were called out the same week.

	Mon	Tue	Wed	Thurs	Fri	Sat	Sun
Programmes	7	6	8	5	7	8	6
Call-outs	2	1	3	0	1	2	2

    a Work out the mean number of comedy programmes each night.
    b A rescue volunteer was asked what the average number of call-outs was per day. What would be the best average to use to answer this?
    c As you can see, as the number of programmes increased so did the call-outs. Draw a scatter graph to confirm this.
    d Does this mean that a higher number of comedies causes more accidents? What does this tell us about making conclusions from this sort of data?

9.4 Creating scatter graphs

1 The table shows the scores students have in their Science and Maths classes.

Student	Jo	Ken	Lim	Tony	Dee	Sam	Pat	Les	Kay	Val	Rod
Science	24	3	13	21	5	15	26	1	12	27	14
Maths	20	5	10	14	9	15	22	7	16	24	11

    a Draw a scatter graph for the data. Use the *x*-axis for Science score, from 0 to 30. Use the *y*-axis for Maths score, from 0 to 30.
    b Describe simply what the graph tells you.
    c Which of the following best describes the degree of correlation?

       positive correlation   negative correlation   no correlation

2 Language students were given oral and written tests. Use a scatter diagram to find out if there is a relationship between the scores from the two types of tests.

Oral test	20	10	14	7	16	8	7	6	14	8
Written test	19	4	8	12	5	12	14	12	10	17

3 In an ice skating competition, two judges gave the following marks for seven skaters. Illustrate their marks on a scatter diagram and comment on the relationship between them.

Judge 1	7.7	8.1	7.9	8.0	9.0	7.4	8.5
Judge 2	6.9	7.9	7.8	8.1	8.9	7.6	7.6

4 Ajit has £16 credit on his mobile phone at the start of the month. The tariff he is on charges 12p per minute. Draw a graph to show how much money is left after every 10 minutes of use.

  a What type of correlation is this?
  b How much credit has he left after 30 minutes?
  c After how minutes is he left with exactly £1?

5 a Draw a scatter graph of the data below. Work out your own scales.

Distance to work (km)	2	5	8	8	9	12	15	16	17	20
Average time taken to get to work (minutes)	6	11	14	29	26	34	37	56	50	55

  b How would you describe the correlation between the time taken to get to work and the distance to work for this group of people: strong, moderate, weak or none at all?
  c Two people live 8 miles away from work. Why do you think one usually gets there much quicker than the other?

6 Sketch a possible scatter graph to illustrate each of the following:

  a the further you drive, the more fuel you use
  b the amount of sunshine and the cost of CDs
  c the higher you climb and the temperature
  d call-outs for mountain rescue teams and number of comedy programmes on TV

Brainteaser

A shop recorded the sales of wellington boots compared to how much snow fell each day.

	Mon	Tues	Wed	Thurs	Fri	Sat
Snowfall in cm	0	3	8	6	2	0
Boots sold	0	1	3	6	4	1

Draw two scatter graphs of this data. On the first, plot the points as normal for each day.

On the second graph, plot Monday's snowfall against Tuesday's sales, then Tuesday's snow against Wednesday's sales and so on.

How does the correlation alter when you plot the sales a day late on the second chart?

Why do you think this is?

10 Algebra

10.1 Algebraic notation

1 Write each of these expressions as simply as possible.

   a $3 \times b$ **b** $p \times 3$ **c** $c \times d$ **d** $5 \times k$

2 Write each of these expressions as simply as possible.

   a $4 \times m \times n$ **b** $u \times 2 \times 6$ **c** $f \times 3 \times t$ **d** $9 \times a \times 4$

   e $r \times r$ **f** $g \times g \times 5$ **g** $h \times \frac{1}{2} \times h$ **h** $7.3 \times p \times p$

3 **a** Write these expressions without using a × sign.

      i $5 \times (q - 3)$ **ii** $(z + 2) \times 9$ **iii** $(8 - 2) \times (e - 2)$ **iv** $(3 - r) \times 10$

    b Write these expressions without using a ÷ sign.

      i $x \div 5$ **ii** $m \div n$ **iii** $y \div 4$ **iv** $35 \div h$

4 Write each of these expressions as simply as possible.

   a $6 - 2 \times t$ **b** $q \times w + 4$ **c** $10 + 5 \times p$ **d** $(4 - 1) \times c + 5$

5 Write each of these expressions as simply as possible.

   a $7 \times 2a$ **b** $3h \times 5$ **c** $0.5 \times 8s$ **d** $6d \times 9$

   e $3w \times 3w$ **f** $7t \times 2t$ **g** $1.5m \times 6m$ **h** $5n \times 4n$

(MR) **6** Rosie has simplified some expressions. State which ones she has answered correctly. Correct her mistakes.

   a $5c \times 3c = 15c$ **b** $4a \times 3b = 12ab$ **c** $g \div 5 = \frac{5}{g}$

   d $2u \times 4v = 6uv$ **e** $8 \times f - 2 = 8f - 2$ **f** $9 - 5 \times y = 4y$

7 Write down the equivalent expressions, for example, $5d = 5 \times d$.

$st + 3$ $3 - st$

$3s + t$

$3 + t \times s$

$3 \div s + t$ $3 + st$

$3 \times s \times t$

$ts - 3$ $s \times 3 + t$

$t + s^3$ $t \times 3 + s$

$\frac{s}{3} + t$ $s^3 t$

$t + \frac{3}{s}$ $t \times s + 3$

$s + 3t$

8 Simplify these expressions.

    a $6n \times 2n$ **b** $5k \times 3k \times 2k$ **c** $h \times 5g \times 4g \times 2 \times 3h$

    d $2u \times 3u \times 5u \times 7u$ **e** $6w \times w \times 9w \times w \times 10w$

    f $a \times a$

10.2 Like terms

1 Make a list of the terms in each of the following.

    a $y + 2x - 3$ **b** $4x = 3 + 2x$ **c** $3x^2 = 14x + 2$

2 Simplify these expressions.

    a $4i + 7i$ **b** $7r + 2r$ **c** $3u + u$

    d $7t - 3t$ **e** $4n - 3n$ **f** $15t - 10t$

    g $3h + 2h + 4h$ **h** $6y + y + 8y$ **i** $m + 5m + 2m$

    j $9p - 5p + 2p$ **k** $8u + 7u - 3u$ **l** $6k - k - 2k$

3 Simplify these expressions.

    a $6d + 4d + 3$ **b** $7 + 3i + 2i$ **c** $10y - 2y + 9$

    d $4p + 2p - 1$ **e** $7 + 7d - 2d$ **f** $6t - 2t + 5u$

    g $9w - 3w + x$ **h** $c + 2c - d$ **i** $5e - e - 4f$

4 Simplify these expressions.

    a $4q + 3q + 6i + i$ **b** $8z + 3z + 4b + 2b$

    c $9u - 3u + 2v + 4v$ **d** $7j - 5j + 3k + 2k$

    e $9m - 5m + 8n - 7n$ **f** $4d + 6d + 9 - 4$

5 Simplify these expressions.

    a $5zt + 4zt + 2as$ **b** $-ab - ab - ab$ **c** $2ad + 8qw + 7ad - 11qw - 10ad + qw$

6 Simplify these expressions.

    a $16k + 4k + 3l$ **b** $7h^2 + 3i - 2i$ **c** $100y + 20x + 30y + 50x$

    d $4p + 7 + 2p - 1$ **e** $2d^2 + 4d - 7d + 2e$ **f** $7abc - 2acb - 3cab - 6bca$

    g $9w^3 + 3w^2 - 10w^3$ **h** $-3fg - 7f - 2fg$ **i** $ab^2 - a^2b$

10.3 Expanding brackets

1 Expand the following brackets.

  a $4(a + 3)$ **b** $3(d + 9)$ **c** $2(3 - s)$ **d** $4(b - 3)$

  e $5(2s + 3)$ **f** $6(4 + 3i)$ **g** $3(3u - 1)$ **h** $6(4 - 5n)$

2 Expand the following brackets.

  a $a(a + 4)$ **b** $b(8 + b)$ **c** $c(c - 2)$ **d** $d(d + 7)$

3 Expand and simplify the following expressions.

  a $6f + 4(f + 2)$ **b** $2k + 3(k + 2)$ **c** $4x + 2(2 + x)$

  d $3(m + 5) + 4m$ **e** $5b + 2(3b + 1)$ **f** $4(2g + 3) + g$

4 Expand and simplify the following expressions.

  a $4(k + 3) + 2(k + 5)$ **b** $7(z - 3) + 5(z + 6)$

  c $9(n - 1) + 4(n - 6)$ **d** $2(y + 12) + 8(y - 3)$

  e $5(p + q) + 2(2p + 3q)$ **f** $4(2i + j) + 3(i - j)$

  g $2(3b - 2a) + 4(b - 2a)$ **h** $5(m - 3n) + 2(3m + 4n)$

5 Write these without brackets.

  a $x(3x + 11)$ **b** $r(5r - 2)$ **c** $j(8 - 7j)$ **d** $f(4f + 7)$

6 Which of these expressions are equivalent?

  a $2(5x + 3) + 2(4x - 1)$ **b** $9(2x - 5) + 3(x + 7)$ **c** $3(5x + 2) + 2(3x - 2)$

  d $7(4x + 1) + 2(3x - 5)$ **e** $4(4x + 3) + 5(x - 2)$ **f** $3(2x + 5) + 4(3x - 4)$

7 Here is a rectangle.

  a Explain why the perimeter of the rectangle is given by $2(3x + 5) + 2(7x - 8)$.

  b Simplify this expression for the perimeter.

  c Find the perimeter of the rectangle when $x = 2$.

$7x - 8$

$3x + 5$

8 Expand and simplify the following expressions.

   a $4t - (9t + 3u)$ **b** $7m - (2m - n)$ **c** $4x - (4y + 5x) - (8x - 3y)$

9 Expand and simplify the following expressions.

   a $4(h + 2i) - 9(2h - 3i)$ **b** $11(2s - 5t) - 8(s - 13t)$

   c $-5(2w^2 - v) - 3(4w^2 + 2v)$

10.4 Using algebraic expressions

1 **a** Write down the price of 3 cups of coffee each costing x pence.

   b Write down the length of a snake which is $2m$ longer than a snake of length L.

   c Write down the amount of change left from £10 if £P has been spent.

   d Work out how many sweets each child gets if T sweets are shared between 4 children.

2 I buy a kettle for £K. Which expression shows the change I get if I pay with a £50 note?

   i $50 + K$ **ii** $50 - K$ **iii** $50K$ **iv** $\frac{50}{K+K}$

3 **i** Write down the perimeter of each rectangle as simply as possible.

   ii Write down the area of each rectangle as simply as possible.

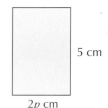
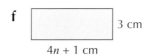

a   3 cm, x cm

b r cm, m cm

c   5 cm, $2p$ cm

d   5 cm, $2x$ cm

f   3 cm, $4n + 1$ cm

4 Write down the perimeter of each shape as simply as possible.

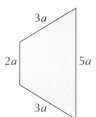
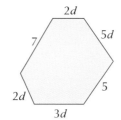
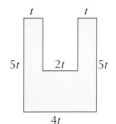

a $3a$, $2a$, $5a$, $3a$

b $2d$, 7, $5d$, $2d$, $3d$, 5

d t, t, $5t$, $2t$, $5t$, $4t$

5 Choose one or more of the weights on the right that, together, will balance the weight on the left.

a

$2x - 1$

$x + 2$

$x + 4$

$2x + 3$

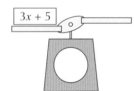

b

$2x + 5$

$2x + 4$

$x + 2$

$3x - 2$

 6 Write down an expression for the volume of each of these cuboids.

a

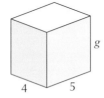

b

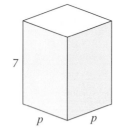

c

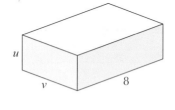

 7 Write down the perimeter and area of each shape as simply as possible.

a

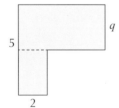

b

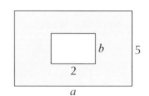

c

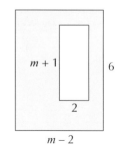

  8 **a** Write down the area of the large parallelogram.

Area of a parallelogram = base × height

b Write down the areas of the smaller parallelograms, A to D.

c Show that the sum of the smaller areas is the same as your answer to part **a**.

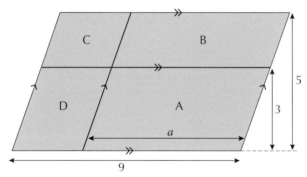

9 For each question, mark any necessary lengths on your diagram.

    a Sketch a rectangle whose area is *abc*.

    b Sketch a rectangle whose perimeter is $6x + 4y$.

    c Sketch a triangle whose area is $8x$.

    d Sketch a rectangle whose area is $5x + 10$.

 10 This is an algebra wall.

Each brick is the sum of the two bricks below it.

Show that the expression in the top brick can be written as $7(x + 2)$.

  11 Show that the expression in the top brick can be written as $4(x - 3)$.

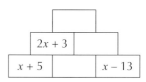

Brainteaser

A Fibonacci sequence is one for which each term (apart from the first two) is found by adding the two previous terms. The original Fibonacci sequence starts with 1, 1 and continues 2, 3, 5, 8, 13, …

The sequences below obey the same rule but start with two different numbers.

Find the missing numbers.

a 7, …, …, 17

b 2, …, …, …, 22

c 5, …, …, …, …, 20

d 4, …, …, …, …, 2

e 3, …, …, …, …, …, …, …, …, …, 354

10.5 Using index notation

 1 Write these expressions as powers (using index form).

    a $4 \times 4 \times 4$ **b** $3 \times 3 \times 3 \times 3 \times 3 \times 3$ **c** $10 \times 10 \times 10 \times 10 \times 10$

 2 Write these expressions as powers (using index form).

    a $a \times a \times a$ **b** $g \times g \times g \times g \times g \times g$

    c $u \times u \times u \times 3$ **d** $4 \times 5 \times t \times t \times t$ **e** $r \times 4r$

    f $3m \times 2m$ **g** $5w \times w \times 2w$ **h** $2j \times 2j \times 2j$

3 Explain the difference between $6w$ and w^6. Write out each in full.

4 Put these expressions in order of size, from smallest to largest, when

    a $u = 4$ **b** $u = 0.5$

    $5u$ u^2 $3u^2$ u^3

5 Simplify each of these expressions.

    a $4b \times a \times a$ **b** $2v \times u \times v \times 3u$

    c $y \times 5z \times 2z \times y$ **d** $G \times H \times 5 \times G \times H \times 5$

6 Write down an expression for the volume of each of these cuboids.

    a   **b**   **c**

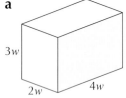
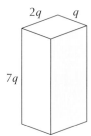
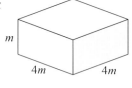

7 Write as simply as possible.

    a $c^2 \times c$ **b** $5c \times 3c^4$ **c** $c^3 \times c^3$ **d** $9c^3 \times 7c^3$

8 Expand the brackets.

    a $4w(5w - w^2)$ **b** $7x^2(x + 2x)$ **c** $8yz(3y^2 - 5z)$

9 Simplify:

    a $2a^2b^3 \times 5ab^5$ **b** $(4c^2)^3$ **c** $(2d^2e)^4$

11 Congruence and scaling

11.1 Congruent shapes

1 Use your ruler to check which of these triangles are congruent. Write your answer like this, in the form P = Q.

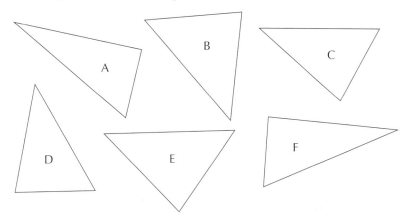

2 Which of these shapes are congruent? Write your answer like this, in the form P = T = W.

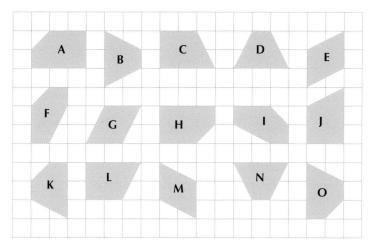

3 a Using triangular dotted paper, draw 14 different shapes by joining some of the dots. Two examples are shown in the diagram below. Label your shapes from A to N.

b Write down the shapes that are congruent in this diagram.

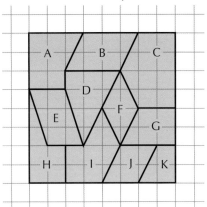

  4 In the diagram each small triangle has an area of 1 unit.

How many congruent triangles are there with:

a area 1 **b** area 4 **c** area 9?

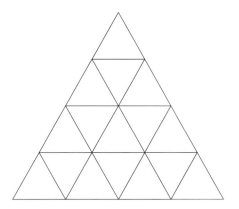

 5 Show how this cross can be split into four congruent pentagons.

6 There are three pairs of congruent triangles pictured below. Find the pairs that go together, and say whether the reason is ASA, SAS or SSS.

A 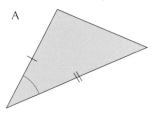 B 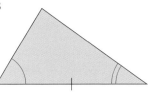 C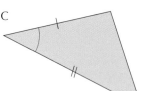

D 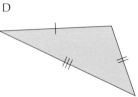 E 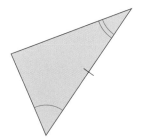 F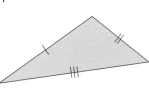

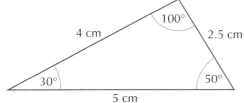

Compare the eight triangles below to the original above. Work out which of these triangles:

a must be congruent to it
b could be congruent, but more information is needed
c are definitely not congruent to it.

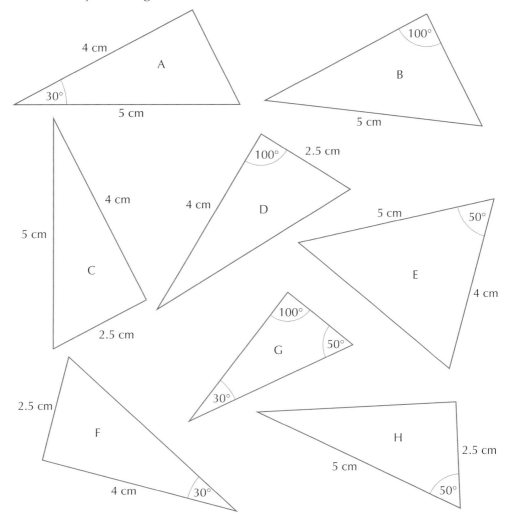

11.2 Enlargements

 1 Trace each shape with its centre of enlargement O. Enlarge the shape by the given scale factor.

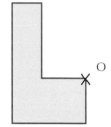

Scale factor 3   Scale factor 2

 2

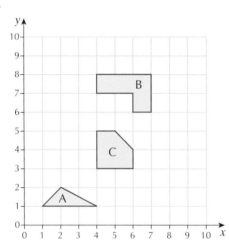

a Copy the grid and shape A only. Enlarge shape A by a scale factor of 2, using the origin as the centre of enlargement. Label the image A'.

b Copy the grid and shape B only. Enlarge shape B by a scale factor of 2, using the point (7, 10) as the centre of enlargement. Label the image B'.

c Copy the grid and shape C only. Enlarge the shape by a scale factor of 3, using the point (5, 4) as the centre of enlargement. Label the image C'.

 3 Triangle ADE is the image of triangle ABC after an enlargement of scale factor 2 from point A.

a Write down the length of AD if AB = 5cm.

b What is the length of CB if ED is 7cm?

c Work out the ratio of the areas of triangle ADE to triangle ABC.

d What is the size of angle ACB if angle CED is 65°?

e Name two congruent triangles.

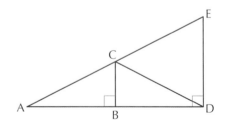

4   Describe fully the two enlargements shown below:

a

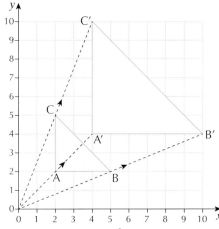

b

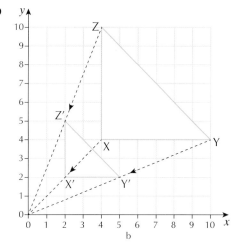

5   **a**   Copy the diagram.
    b   Shape B is an enlargement of shape A.
            What is the scale factor?
    c   Draw ray lines to find the centre of
            enlargement. Write down the coordinates
            of this point.

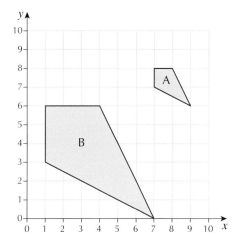

MR 6   Using an *x*-axis from 0 to 9 and *y*-axis from 0 to 6, plot the following points:

A(0,0) B(0,1) C(1.5, 2) D(3,1) E(3,0)

a   What is the mathematical name for the shape you get?
b   Enlarge the shape with scale factor 2, centred on the origin. Label the new points
        A′, B′, etc.
c   Do the same with scale factor 3. Label these new points A″, B″, etc.
d   Which point does not move its position? Does this happen with any enlargement?
e   Write the new coordinates in a table like the one shown:

Original shape	Enlargement scale factor 2	Enlargement scale factor 3
A (0, 0)	A′ ( )	A″( )
B (0, 1)	B′	
C (1.5, 2)	C′	
D (3, 1)	D′	
E (3, 0)	E′	

11.3 Shape and ratio

1 Express each of these ratios in its simplest form.

a   90 cm : 20 cm   b   32 mm : 72 mm

c   150 cm : 2 m   d   1.8 cm : 40 mm

e   0.65 km : 800 m   f   45 cm to 3 m

2 The land area to sea area of the surface of the Earth is roughly in the ratio 3 : 7.
The total surface area of the Earth is 500 000 000 km². What is the area of land?

3 Which of these shapes are in the same ratio? (They are not drawn to scale.)

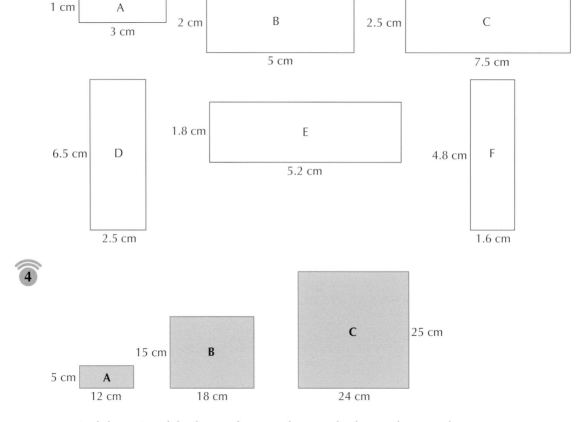

4

a   Find the ratio of the base of rectangle A to the base of rectangle B.
b   Find the ratio of the area of rectangle A to the area of rectangle B.
c   What fraction is area A of area B?
d   Find the ratio of the area of rectangle B to the area of rectangle C.
e   Which is greater: area A as a fraction of area B, or area B as a fraction of area C?

5 The diagram shows the design of a new flag.

   a Calculate the ratio of the red area to the green area.

   Use ratios to answer the following questions.

   b 60 m² of red cloth is used to make some flags. How much green cloth is needed?

   c The total area of some flags is 200 m².

   i How much red cloth do they contain?

   ii How much green cloth do they contain?

   iii If red cloth costs £8 per square metre, and green cloth costs twice as much, how much would it cost to make these flags?

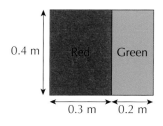

 6 The diagram shows the plan of a garden.

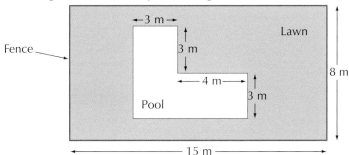

   a Find the ratio of the perimeter of the fence to the perimeter of the pool.
   b Calculate the area of the pool.
   c Calculate the area of the lawn.
   d Find the ratio of the lawn area to the pool area.

 7 A school has lots of small dice that are cubes measuring 2 cm along each side, but a teacher wants to use a large foam dice to demonstrate. This dice has sides measuring 10 cm.

   a Work out the ratio of the sides of the two cubes. Give your answer in the form 1 : *n*.
   b Work out the ratio of the total surface area of the two cubes. Give your answer in the form 1 : *n*.
   c Work out the ratio of the volume of the two cubes. Give your answer in the form 1 : *n*.
   d The ratio of the sides of two other cubes is 1 : 4.

   i Write down the ratio of the total surface area of the two cubes.
   ii Write down the ratio of the volume of the two cubes.

In the rainbow triangle, the base of each successive coloured triangle increases by the length of the violet triangle's base.

a What is the ratio of these base lengths?

   i violet to green   **ii** red to violet

   iii blue to yellow

b What is the ratio of these areas?

   i violet to blue   **ii** orange to violet

   iii green to red

c What do you notice about the areas when you compare *red and violet* to *orange and blue* to *yellow and green?*

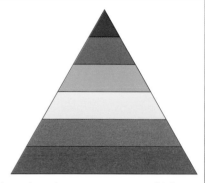

11.4 Scales

1 These objects have been drawn using the scales shown. Find the true lengths of the objects. (In part **a**, measure the length of the golf club shaft only.)

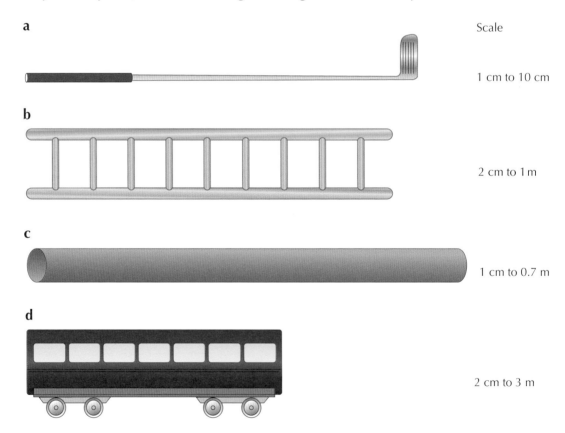

a

Scale

1 cm to 10 cm

b

2 cm to 1 m

c

1 cm to 0.7 m

d

2 cm to 3 m

2 The objects are shown with their real dimensions. Work out their size on a scale drawing using the scales shown.

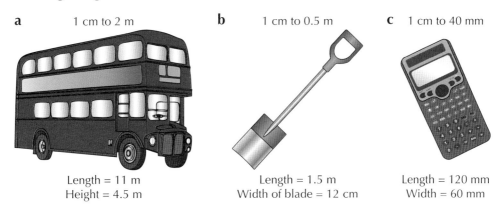

a 1 cm to 2 m

Length = 11 m
Height = 4.5 m

b 1 cm to 0.5 m

Length = 1.5 m
Width of blade = 12 cm

c 1 cm to 40 mm

Length = 120 mm
Width = 60 mm

3 The diagram shows a scale drawing of an aircraft hangar (plan view).

Scale: 1 cm to 120 m

a Calculate the real length of the hangar.
b Calculate the real width of the hangar.
c Calculate the area of the hangar.

4 Copy and complete the table.

	Scale	Scaled length	Actual length
a	1 cm to 3 m	5 cm	15 m
b	1 cm to 2 m		24 m
c	1 cm to 5 km	9.2 cm	
d		6 cm	42 miles
e	5 cm to 8 m	30 cm	

5 The diagram shows a scale drawing of a shop, where 1 cm represents 2 m. Make a table showing the real-life dimensions and area of each section of the shop.

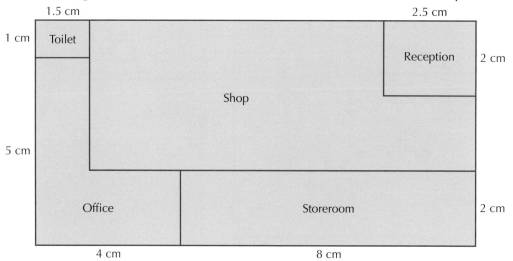

a Work out **i** the scaled area and **ii** the real-life area of each part of the shop.
b **i** Write the scale factor of the dimensions as a ratio in the form 1 : *n*.
  ii Write the scale factor of the areas as a ratio in the form 1 : *n*.

6

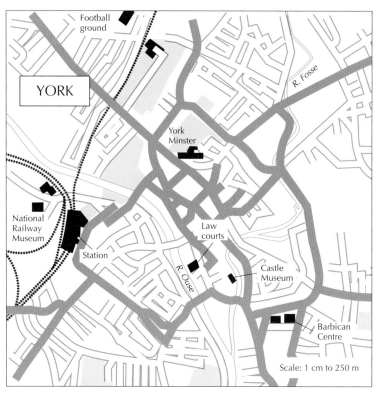

a Write the scale in map format.
b How far is it from the Minster to **i** the Barbican **ii** the Railway Museum?
c Some football fans need to get catch a train at 5:30 pm to get home. If the match finishes at 4:55 and their fastest walking pace is 8 m/s, how much time will they have to spare? (Remember they cannot go direct as the black line is the railway, so you will have to find their shortest route by road.)

d In summer, boat trips go between the two bridges over the River Ouse, marked in red. If the return trip takes 30 minutes, what is the average speed of the boat in
  i km/h   **ii** m/s?

e The boat trip costs £3.50 per adult and £2.50 per child. Mr Ahmed takes his wife and two children.
  i   How much does it cost him?
  ii He thinks any activity that costs less than 50p a minute is good value. Is this trip good value in his opinion?

f What is the approximate area of the railway station?

g The Castle Museum currently measures 2 mm by 3 mm on the map. What are the real-life dimensions and area? They want to double their size, and the curator thinks doubling both of these dimensions will do that. Is she right? What will actually happen if they do this?

12 Fractions and decimals

12.1 Adding and subtracting fractions

1 Write each of the following as a mixed number.

   a $\dfrac{7}{4}$ **b** $\dfrac{5}{2}$ **c** $\dfrac{9}{6}$ **d** $\dfrac{13}{2}$ **e** $\dfrac{11}{3}$ **f** $\dfrac{8}{7}$

2 Add these fractions. Give your answer as simply as possible (as a mixed number if necessary).

   a $\dfrac{3}{7} + \dfrac{2}{7}$ **b** $\dfrac{5}{9} + \dfrac{1}{9}$ **c** $\dfrac{3}{10} + \dfrac{9}{10} + \dfrac{7}{10}$

3 Subtract these fractions. Give your answer as simply as possible.

   a $\dfrac{4}{5} - \dfrac{1}{5}$ **b** $\dfrac{8}{9} - \dfrac{5}{9}$ **c** $\dfrac{7}{8} - \dfrac{3}{8}$

4 Add these fractions. Give your answer as simply as possible (as a mixed number if necessary).

   a $\dfrac{2}{5} + \dfrac{1}{2}$ **b** $\dfrac{5}{8} + \dfrac{1}{12}$ **c** $\dfrac{5}{6} + \dfrac{1}{3}$ **d** $\dfrac{4}{7} + \dfrac{3}{5}$

5 Subtract these fractions. Give your answer as simply as possible.

   a $\dfrac{3}{5} - \dfrac{1}{2}$ **b** $\dfrac{5}{9} - \dfrac{1}{6}$ **c** $\dfrac{7}{8} - \dfrac{2}{3}$ **d** $\dfrac{7}{10} - \dfrac{1}{4}$

6 Calculate these.

   a $\dfrac{3}{7} + \dfrac{2}{3}$ **b** $\dfrac{5}{12} + \dfrac{1}{8}$ **c** $\dfrac{7}{9} + \dfrac{5}{6}$ **d** $\dfrac{3}{5} + \dfrac{1}{2} + \dfrac{7}{10}$

   e $\dfrac{7}{9} - \dfrac{1}{2}$ **f** $\dfrac{11}{15} - \dfrac{2}{5}$ **g** $\dfrac{5}{6} - \dfrac{1}{10}$ **h** $\dfrac{2}{3} + \dfrac{5}{6} - \dfrac{5}{12}$

   i $\dfrac{11}{18} - \dfrac{7}{24}$ **j** $\dfrac{19}{40} - \dfrac{43}{60}$

7 Calculate these.

   a $2\frac{1}{5} - \frac{7}{10}$ **b** $1\frac{3}{4} + \frac{5}{6}$ **c** $3\frac{1}{8} - 2\frac{2}{3}$ **d** $4\frac{1}{10} - 2\frac{1}{4}$

   e $2\frac{3}{25} + 4\frac{7}{10}$ **f** $8\frac{3}{16} - 5\frac{11}{12}$

8 Of Ian's emails, $\frac{3}{5}$ is junk mail and $\frac{1}{10}$ is from friends. The rest is work-related.

    a What fraction is work-related?

    b Ian received 120 emails during the week. How many were not junk mail?

  9 Sharon bought a bag of flour weighing $3\frac{3}{4}$ kg. She used $\frac{7}{10}$ kg for a cake and $1\frac{3}{5}$ kg for some bread.

    a How much flour did she use altogether?

    b How much flour did she have left over?

  10 Work out each answer.

    a $\dfrac{5}{12} + \dfrac{5}{24} + \dfrac{5}{36}$ **b** $\dfrac{9}{10} + \dfrac{7}{8} + \dfrac{5}{6} + \dfrac{3}{4} + \dfrac{1}{2}$

12.2 Multiplying with fractions and integers

 1 Work out:

    a $\dfrac{1}{4}$ of 28   **b** $\dfrac{3}{5}$ of 30   **c** $\dfrac{1}{3}$ of 36   **d** $\dfrac{3}{4}$ of 24

 2 Calculate each of these.

    a $\dfrac{4}{9}$ of 36 kg   **b** $\dfrac{5}{6}$ of 30 ml   **c** $\dfrac{2}{7}$ of 28 cm   **d** $\dfrac{1}{4}$ of 52 km

    e $\dfrac{3}{7}$ of 42 cm   **f** $\dfrac{7}{10}$ of 50 g   **g** $\dfrac{3}{5}$ of £80   **h** $\dfrac{7}{8}$ of 640 litres

 3 Calculate each of these. Cancel your answers and write them as mixed numbers if necessary.

    a $\dfrac{2}{3} \times 6$ **b** $\dfrac{3}{7} \times 4$ **c** $\dfrac{4}{5} \times 7$

    d $\dfrac{2}{9} \times 8$ **e** $\dfrac{3}{8} \times 4$

4 Work out the areas of these rectangles.

a
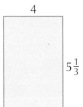
4
$5\frac{1}{3}$

b
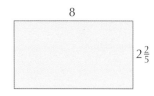
8
$2\frac{2}{5}$

c
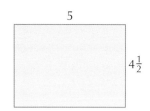
5
$4\frac{1}{2}$

 5 Work out:

a $5\frac{1}{6} \times 3$ **b** $3\frac{3}{4} \times 5$ **c** $6\frac{1}{2} \times 8$ **d** $3\frac{4}{5} \times 2$

 6 $42\frac{6}{7}\% = \frac{3}{7}$

Use this fact to work out:

a $42\frac{6}{7}\%$ of 14   **b** $42\frac{6}{7}\%$ of 35   **c** $42\frac{6}{7}\%$ of 560

d $42\frac{6}{7}\%$ of 9800   **e** $21\frac{3}{7}\%$ of 420   **f** $142\frac{6}{7}\%$ of 91

  7 Sylvia has £660.

Sylvia gives $\frac{2}{3}$ of the money to Dabira.

Dabira gives $\frac{4}{5}$ of the money she receives to Kirsty.

Kirsty gives $\frac{7}{8}$ of the money she receives to Yasmin.

Yasmin gives $\frac{3}{4}$ of the money she receives to Titomi.

Titomi gives $\frac{6}{11}$ of the money she receives to Sylvia.

How much does each person have at the end?

Brainteaser

A company was contracted to paint the walls and roof of a cuboid building.

The base measured 16 m by 13 m and the building was 5 m tall.

The building had 18 m² of windows.

Sivas painted $\frac{1}{12}$ of the building.

Elakiya painted $\frac{4}{15}$ of the building.

Ire painted $\frac{3}{20}$ of the building.

Sabah painted $\frac{5}{16}$ of the building.

Esosa had 20 cans of paint. Each can was enough to paint 4 m².

Did Esosa have enough paint to finish the job?

12.3 Dividing with fractions and integers

 1   **a**   What is one-half of $\frac{1}{3}$?

     b   What is one-third of $\frac{1}{3}$?

     c   What is one-fifth of $\frac{1}{3}$?

 2   Work out:

   a $\frac{3}{4} \div 6$ **b** $\frac{3}{4} \div 10$ **c** $\frac{3}{4} \div 9$ **d** $\frac{3}{4} \div 15$

 3   Work out:

   a $\frac{2}{5} \div 3$ **b** $\frac{8}{9} \div 6$ **c** $\frac{2}{3} \div 6$ **d** $\frac{3}{7} \div 4$

   e $\frac{1}{11} \div 5$ **f** $\frac{4}{5} \div 8$ **g** $\frac{5}{7} \div 20$ **h** $\frac{2}{3} \div 8$

 4   Work out:

   a $2\frac{1}{2} \div 6$ **b** $5\frac{2}{5} \div 4$ **c** $4\frac{3}{4} \div 8$ **d** $9\frac{4}{9} \div 10$

  5   Sharbini has $10\frac{1}{2}$ kg of wheat to share between seven people.

   How much wheat will each person get?

 6   **a**   The perimeter of a pentagon is $7\frac{3}{4}$ cm. Work out the length of each side.

     b   The perimeter of an octagon is $12\frac{2}{5}$ cm. Work out the length of each side.

 7   Kailash measured the length and breadth of this picture frame that he made.

   a   What is the total length of frame material used?

   Kailash has 50 inches of frame material to make an identical frame.

   b   How much frame material will be left over?

   He bought a 97-inch strip of frame material.

   c   How many of each frame could he make from it? Show your working.

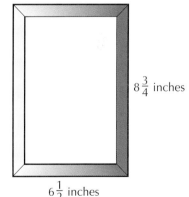

$8\frac{3}{4}$ inches

$6\frac{1}{2}$ inches

 8   Naiga made three cakes and then ate $\frac{5}{8}$ of one.

   Amy made two cakes and then ate $\frac{5}{8}$ of one.

   The rest was shared equally among 12 people.

   What fraction of a cake did each of these people receive?

12.4 Multiplication with large and small numbers

 Work out:

a 70×5 b 40×30 c 200×9

d 800×60 e 5000×3 f 30×700

 Work out:

a 0.6×7 b 0.9×9 c 0.5×5 d 0.2×3

 Work out:

a 0.6×0.7 b 0.9×0.9 c 0.5×0.7 d 0.2×0.3

 Work out:

a 0.3×0.04 b 0.09×0.4 c 0.06×0.06 d 0.007×0.2

 Work out:

a 40×0.7 b 0.4×60 c 20×0.2

d 0.9×90 e 0.08×300 f 9000×0.004

g 70×0.04 h 300×0.002

 A seed weighs 0.04 g. How much do 600 seeds weigh?

 Calculate the following.

a $300 \times 0.4 \times 0.8$ b $0.006 \times 7000 \times 0.2$ c $0.04 \times 0.07 \times 300$

 Sound travels about 0.3 km in 1 second. How far does sound travel in the following times?

a 200 seconds   b 10 minutes   c 0.02 seconds

Work in kilometres.

 $88^2 = 7744$

Use this fact to work out:

a 8800^2 b 0.88^2 c 8.8^2 d 0.088^2

10 A rectangle measures 60 cm by 80 cm.
   a What is the area of the rectangle in square centimetres?
   b Find the area of the rectangle in square millimetres.
   c Find the area of the rectangle in square metres.
   d Find the area of the rectangle in square kilometres.

 11 A cuboid measures 30 cm by 40 cm by 50 cm.
   a Find the volume of the cuboid in cubic metres.
   b Find the volume of the cuboid in cubic kilometres.

Brainteaser

Two friends, Abeola and Kirthana, both play five different games online. Each game is either won or lost.

The probability that Abeola wins the first game is 0.2 and for Kirthana 0.3.

For the second game, they both have a probability of winning of 80%.

The probability that Abeola wins the third game is $\frac{1}{5}$ chance. The probability that Kirthana wins the third game is a quarter of Abeola's.

For the fourth game, Abeola's probability of victory is 0.35. Kirthana is twice as likely to win the fourth game as Abeola.

For the fifth and final game, the probability that Abeola loses is 0.6. The probability that Kirthana loses is $\frac{3}{20}$.

Abeola and Kirthana play each game 200 times, except Kirthana who only plays the second game 150 times.

Who is likely to win more games?

12.5 Division with large and small numbers

 1 Calculate these.

   a $9 \div 0.3$ **b** $60 \div 0.3$ **c** $500 \div 0.2$ **d** $5000 \div 0.5$

 2 Calculate these.

   a $30 \div 0.6$ **b** $600 \div 0.3$ **c** $20 \div 0.05$
   d $3000 \div 0.1$ **e** $80 \div 0.02$

3 £1 buys 0.05 g of platinum. How much does 4 g of platinum cost?

  4 The table shows the postage needed for some letters.

Format	Weight	First-class price	Second-class price
Letter	0–100 g	£0.34	£0.24
Large letter	0–100 g	£0.48	£0.36
Large letter	101–250 g	£0.70	£0.53

Answer these questions by working in decimals.

a What is the cost of sending 14 letters each weighing 90 g by first-class mail?
b How many second-class 45 g letters could be posted for £4.80?
c Which is cheaper to post: 18 first-class letters each weighing 130 g or 26 second-class letters each weighing 85 g?

 5 Work out:

a $0.6 \div 3$ **b** $0.8 \div 2$ **c** $0.36 \div 2$ **d** $0.9 \div 3$

e $0.4 \div 1$ **f** $0.15 \div 3$ **g** $0.35 \div 5$ **h** $0.08 \div 2$

 6 Calculate these.

a $35.2 \div 1.1$ **b** $11.7 \div 2.6$ **c** $3.12 \div 4.8$ **d** $13.23 \div 2.7$

 7 Work out:

a $40 \div 0.5$ **b** $4000 \div 0.05$ **c** $0.04 \div 0.05$ **d** $0.0004 \div 5$

  8 An advertiser pays 0.004p to a website every time it is hit (every time the website receives a visit). In one month, the advertiser pays the website £16. How many hits did the website receive?

  9 A website gives 3.6p to charity on every sale it makes. In September, the website gave a mean average of £2.80 per day to charity. How many sales did it make in September?

 10 **a** 3000 bacteria have a mass of 0.000 003 g. What is the mass of twelve bacteria?
b 10 000 000 000 000 atoms have the same mass as one bacterium. What is the mass of 19 atoms? (Bacteria is the plural of bacterium.)

13 Proportion

13.1 Direct Proportion

1 2 gallons is approximately 9 litres.

    a How many gallons are equivalent to 63 litres?
    b How many litres are equivalent to 7 gallons?

FS **2** **a** 8 light bulbs cost £7.20. How much does a box of 20 cost?
    b How many bulbs can you buy for £9?

3 9 grams of silver are used to make 15 cm of chain.

    a How much silver does 20 cm of chain contain?
    b How long is a chain that contains 27 grams of silver?

4 2 of the carriages on a train are first class. The other 6 carriages are second class.

    a What proportion of carriages are first class, as a fraction?
    b What is the ratio of first-class to second-class carriages?
    c A train with 12 carriages has the same proportion of first-class carriages. How many first-class carriages does it have?

5 The perimeter of a circle is called the circumference. The circumference of a circle is proportional to the diameter of the circle. A circle with a diameter of 7 m has a circumference of 22 m.

    a What is the scale factor as a decimal, to two decimal places?
    b Work out the circumference of a circle with a diameter of: **i** 21 cm   **ii** 30 m.
    c Work out the diameter of a circle whose circumference is: **i** 44 mm   **ii** 55 m.

PS **6** A car uses 20 litres of petrol in travelling 140 km.

    a How much would be used on a journey of: **i** 35 km   **ii** 210 km.
    b A full tank is 60 litres. At the end of a journey there was 15 litres left in the car. How long was the journey?

MR **7** The ratio of nitrogen to oxygen in the air is approximately 4 : 1. A cupboard contains 600 litres of air.

    a How much nitrogen and oxygen does it contain?
    b What percentage is oxygen?
    c How much nitrogen and oxygen will a 900 litre cupboard contain?
    d What percentage of this cupboard is oxygen?
    e What does this tell you about the proportion of gases as the cupboard size increases?

8 12 kg of a new metal alloy was made using this formula:

    a tin 400 g   **b** copper 3 kg   **c** lead 1800 g   **d** iron 6.8 kg.

Calculate how much of each component is needed to make 21 kg of the metal alloy.

13.2   Graphs and direct proportion

1 The number of faces on a group of dice can be found from the formula $y = 6x$.
Copy and complete the table for various numbers of dice.

Number of dice	1	2	3	4	5
Number of faces					

Draw a graph of these results with 0 to 5 on the x-axis and 0 to 30 on the y-axis.

2 A helicopter is travelling at a constant speed. The distance travelled is proportional to the time taken. It travels 7 km every 5 minutes.

Time taken (minutes)	5	10	15	20	25
Distance (km)	7				

   a Copy and complete this table.
   b What is the scale factor as a decimal?
   c Write down the formula for distance, D, compared to time, T.

3 Some perimeters (y cm) of a rhombus of side x cm are shown.

Side (x cm)		4		9	
Perimeter (y cm)	10	16	24	36	60

   a Copy and complete this table to show values of x and y.
   b Find the formula.

4 1 inch is almost exactly 2.5 centimetres.

   a What is the formula to represent this, with x as inches and y as centimetres?
   b Draw a table to show the conversion from inches to centimetres for the first 6 inches.
   c Draw a graph of this conversion.

5 This graph shows the exchange rate between pounds and Canadian dollars. £1 = $1.8

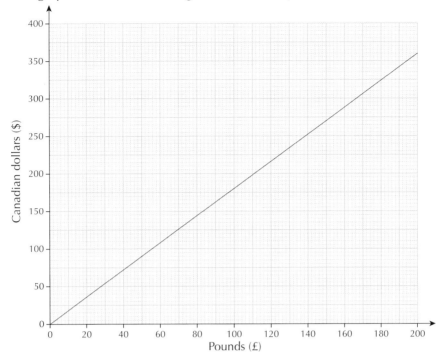

a Mark visits his sister in Toronto and needs some money for when he gets there. How many dollars will he get for £150?

b He has $36 left at the end of his visit. How many pounds will he get back?

c What is the formula connecting the two currencies, using p for pounds and d for dollars?

6 The graph shows how much diesel fuel is used by a car on a 300 mile journey.

Copy the table shown and fill in any gaps.

Distance (x miles)		30		96	
Diesel (y litres)	2		6		12

a What is the formula linking diesel usage with miles travelled?

b How much diesel would be used up after 60 miles?

c The car began the journey with 20 litres in the tank. Will the driver have to refuel before he finishes the journey?

d With diesel priced at £1.40 per litre, how much will the journey have cost?

e A petrol car uses fuel with a formula of $y = 10x$, but with petrol only costing £1.30 per litre, which type of fuel works out cheapest for this journey, and by how much?

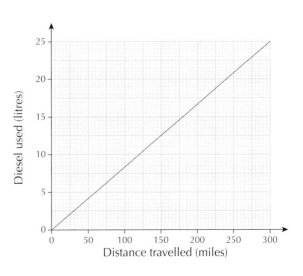

13.2 Graphs and direct proportion   **93**

Brainteaser

A well-known make of margarine contains 70 g of fat in every 100 g of its product.

a Copy and complete this table to show how much fat the margarine contains for the given amounts and draw a graph labelled A.

Amount of margarine (g)	100	200	300	400	500
Amount of fat (g)					

b What percentage of the margarine is fat?

Another brand claims to have 20% less fat than the first brand. It does not make clear whether this means 20% of the margarine (B) or 20% of the fat (C).

c Draw new tables for each possibility and then draw their graphs (perhaps in different colours) on the same axes as your first graph.

d Which of the three graphs shows the healthiest option? What do you notice about its slope?

e What is the difference in the three amounts of fat at 500 g?

13.3 Inverse proportion

1 4 pipes can fill a tank in 70 minutes. How long will it take to fill the tank by 7 equal-sized pipes?

2 It takes 8 people 12 hours to dig a hole. How long would it take the following numbers of people?

    a 4 people   **b** 2 people   **c** 6 people

3 Imran bought 40 toys at £12 each. How many toys can Imran buy at £8 each if he spends the same amount in total?

4 The table shows the time taken to run 100 metres for various running speeds.

Speed (metres/second)	4	5	8	10	20
Time taken (seconds)	25	20	12.33	10	5

    a Most Olympic athletes can now run 100 metres in less than 10 seconds which is 10 metres/second. What is 10 metres/second in kilometres per hour?

    b To the nearest whole number, how fast would you have to run to get below 9 seconds? A graph may help, but try to find coordinates for 6, 7 and 9 seconds to get an accurate curve.

    c A PE teacher reckons he can run 100 metres in 15 seconds. How fast does he need to run?

5 A village held a sponsored 10 mile road race. The graph shows how long it took to complete it depending on people's average speed.

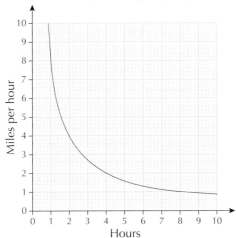

a The winner finished after 55 minutes. What was his average speed to one decimal place?

b The youngest entrant averaged 2 mph. How long did it take her to finish?

c The run started at 11 am. About what time did someone finish whose average speed was 6 mph?

d A disabled army veteran walked the entire course on crutches at an average speed of 1.5 mph, but needed a 5-minute break after each hour. Did he finish before or after 5 pm?

6 y is inversely proportional to x. When $y = 3$, $x = 12$.

a What does xy equal?

b What will x be when $y = 4$?

c What will y be when $x = 1$?

d What other whole number combinations are possible?

7 Meryl wants to make a rectangular garden bed with an area of 24 m².

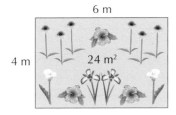
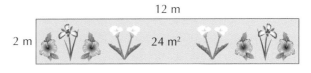

a Draw up a table to show these and other possible length and width measurements of the garden bed.

b Do the measurements have to be whole numbers? Explain your answer.

c Draw a graph to show the relationship between the length and the width of the garden bed.

d In order to sum up what happens to the area, fill in these blanks:
As the length of one side _____, the _____ has to be halved for the _____ to stay the same.

8 When $x = 1.6$, $y = 16$, but when $x = 4$, $y = 6.4$.

    a Work out the formula for this relationship.

    b Draw up a table for integer values of x from 1 to 8. Round your answer to three decimal places.

    c Draw a graph that allows answers to be read accurately to one decimal place.

    d What is y when $x = 3\frac{1}{3}$

Brainteaser

Leo decides to get a group of friends together into a syndicate to play a lucky draw game similar to a lottery. The bigger the group, the more tickets they can buy, but if they win, they will have to share the prize among more people. The total amount of prize money is £3000.

a How will the number in the group affect the amount each person receives?

b What kind of relationship is this?

c Copy and complete the table below.

Number of people	1	2	4		8	10		20	
Share of money (£)	3000	1500		500			200		60

d Plot a graph of these points to show the relationship.

e How much would 40 people get each?

f What is the maximum number of people that can join if they want to win at least £100?

g Why would it be difficult to share the prize money if 23 people joined the syndicate?

13.4 Comparing direct proportion and inverse proportion

    1 Which of these graphs shows **a** direct proportion and **b** inverse proportion?

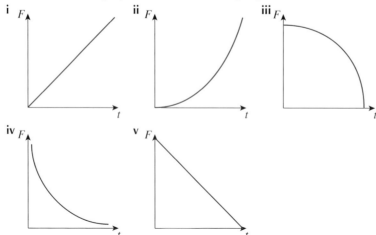

2 Sketch a graph to show what each of these might look like.

a distance travelled and time taken

b number of workers doing a repair and time taken to complete the job

c number of texts on a pay-as-you-go mobile phone and amount of credit remaining

d depth of bath water and time taken for it to empty

3 Decide which of these you think is direct or inverse proportion:

a distance travelled and fuel used

b length of dress and material used

c running speed and time taken to complete a race

d number of windows in a house and time taken to clean them

e diameter of a tube and time taken for water to get through

f hours worked and amount in pay packet

4 In the table below, x and y are directly proportional.

You can always write $y = mx$ where m is a number. In this example $y = 8x$.

Copy and complete the table.

x	3	4.5	6		20
y	24		48	120	160

5 What does y equal for each of these?

a

x	1	2	3	4	5
y	13	26	39	52	65

b

x	2	4	6	8	10
y	7	14	21	28	35

c

x	5	10	15	20	25
y	20	40	60	80	100

6 In this table, x and y are inversely proportional. You can always write $xy = k$ where k is a number.

x	3	4.5	6	15	
y	24	16		4.8	3.6

This time $xy = 72$, and the formula could be rearranged to $y = \frac{72}{x}$

Check that each pair of numbers does multiply to make 72, then copy and complete the table.

Draw a graph from the table.

7 What is the formula for these tables?

a

x	1	2	3	4	5
y	60	30	20	15	12

b

x	1	3	5	7.5
y	75	15	25	10

c

x	2	3	4	6	8
y	48	32	24	16	12

 8 Decide whether each of these represents direct proportion, inverse proportion or neither.

Draw the graph for each and write down a formula where possible.

a

x	1	2	3	4.2	6
y	42	21	14	10	7

b

x	1	2	3	4	5
y	13	26	39	52	65

c

x	10	8	6	4	2
y	5	4	3	2	1

d

x	12	40	15	2	5
y	10	3	8	60	24

14 Circles

14.1 The circle and its parts

1   a   A circle has a radius of 6 cm. What is the length of its diameter?

   b   A circle has a diameter of 30 m. What is the length of its radius?

2   a   Draw a circle with radius 33 mm.

   b   Draw a circle with diameter 9.2 cm.

   c   Draw a semicircle with diameter 8 cm.

   d   Draw a quadrant of a circle with radius 52 mm.

3   Construct these diagrams.

   (Hint: To find out the radius of each circle, measure the diameter, and halve it.)

   a   b

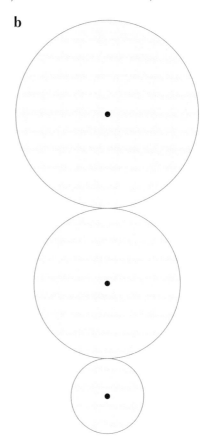

4   **a**   Name the lines on this diagram.

  i   AB   **ii**   AC   **iii**   DE

  iv   CF   **v**   CG

  b   Name these shapes.

  i   ABCA   **ii**   ACFA

  iii   ABCFA   **iv**   GABCG

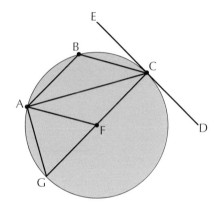

5   Construct these diagrams.

  a

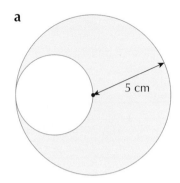

5 cm

  b

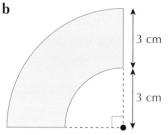

3 cm

3 cm

  c

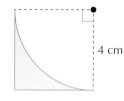

4 cm

14.2 Formula for the circumference of a circle

1   For this question, take π = 3.

Find the circumference of a circle with:

  a   a diameter of 10 m

  b   a radius of 4 m

For the rest of this exercise, take π = 3.14 or use the π key on your calculator.

2   Calculate the circumference of each wheel.

Write your answers correct to the nearest centimetre.

  a

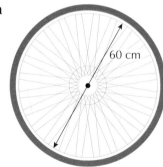

60 cm

  b

17.2 cm

3 Calculate the circumference of each coin.

Give each answer to one decimal place.

a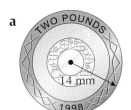

b 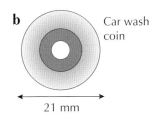 Car wash coin

21 mm

4 The world's tallest ferris wheel is the Singapore Flyer, opened in 2008.

It has a diameter of 150 m.

It has 28 capsules, spread equally around the circumference.

Kamile gets into a capsule. Ewa gets into the next capsule.

Find the distance travelled by Kamile before Ewa gets into her capsule, correct to the nearest metre.

5 Calculate the total length of the lines in this crop circle diagram.

It has two semicircles, a circle and straight lines.

Write your answers correct to the nearest centimetre.

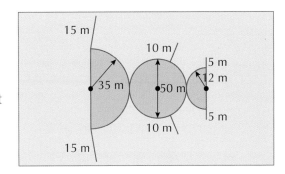

6 Calculate the total perimeter of this arched window.

Give the answer correct to the nearest centimetre.

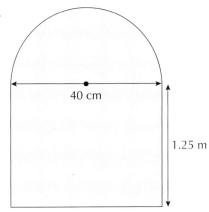

40 cm

1.25 m

7 The radius of this quadrant is 20 mm.

Calculate the perimeter of this quadrant in millimetres.

Give your answer to three significant figures.

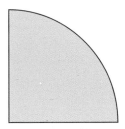

8 Calculate the total perimeter of this shape.
Give your answer correct to two decimal places.

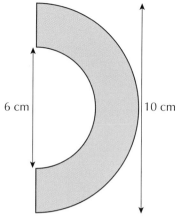

6 cm | 10 cm

9 Calculate the diameter of a circle with a circumference of 500 cm.
Write your answer correct to the nearest centimetre.

10 The circumferences of four planets are given in the table.
Find the radius of each one, correct to the nearest kilometre.

Jupiter	439 264 km
Saturn	365 882 km
Uranus	159 354 km
Neptune	154 705 km

14.3 Formula for the area of a circle

1 For this question, take $\pi = 3$.
Find the area of a circle with:

a a diameter of 10 m
b a radius of 4 m

For the rest of this exercise, take $\pi = 3.14$ or use the π key on your calculator.

2 Calculate the area of each of the following circles. Give each answer to one decimal place.

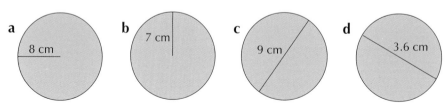

a 8 cm   b 7 cm   c 9 cm   d 3.6 cm

3 Calculate the area of the lid of each tin. Write your answers correct to one decimal place.

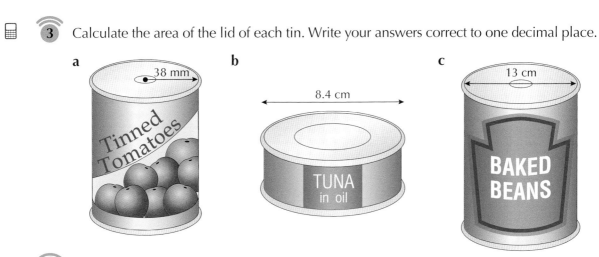

a
38 mm
Tinned Tomatoes

b
8.4 cm
TUNA in oil

c
13 cm
BAKED BEANS

4 A CD has a diameter of 12 cm.
Calculate its circumference and area. Give your answers to one decimal place.

5 Calculate the total area of this arched window.
Write your answer correct to the nearest square centimetre.

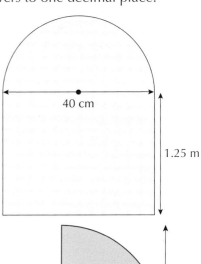

40 cm
1.25 m

6 Calculate the area of this shape.
Write your answer correct to the nearest square centimetre.

6 cm   10 cm

7 Find the area of this shape.
Give your answer in terms of π.

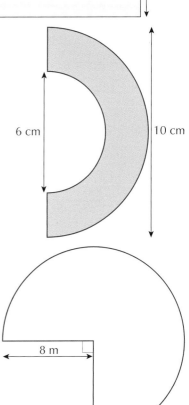

8 m

8 This diagram shows how a washer is made.

a Calculate the area of the blank, correct to one decimal place.

b Calculate the area of the hole, correct to one decimal place.

c Calculate the area of the finished washer, correct to the nearest square millimetre.

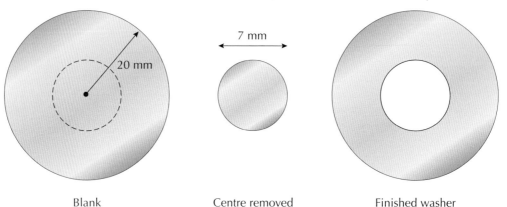

| Blank | Centre removed | Finished washer |

9 The minute hand on a clock has a length of 24 cm.

Calculate the area swept by the minute hand in:

a 1 hour

b 5 minutes

c 1 minute.

Give your answers in terms of π.

Brainteaser

For this brainteaser, use π = 3.

Find the value of H.

A is the diameter of a circle with a radius of G.

B is the circumference of a circle with a diameter of D.

C is the area of a semicircle with a diameter of F.

D is the radius of a quadrant with an area of 48 cm².

E is the perimeter of a semicircle with a diameter of B.

F is the area of a circle with a diameter of (B − A).

G is the radius of a circle with a circumference of E.

H is the perimeter of a quadrant with a radius of C.

15 Equations and formulae

15.1 Equations with brackets

 1 Solve the following equations.

    a $4d + 12 = 32$ **b** $4t - 8 = 12$ **c** $10s + 5 = 35$

    d $6g + 30 = 60$ **e** $6f - 3 = 9$ **f** $12q + 15 = 51$

 2 Solve the following equations.

    a $4(s + 2) = 4$ **b** $3(m - 5) = -6$ **c** $2(3n + 9) = 6$

    d $5(2y - 12) = -10$ **e** $-5x = 10$ **f** $8 - 5x = -12$

    g $12 + 5f = -3$ **h** $2(x + 3) = -10$ **i** $3w - 7 = -1$

    j $4(2d - 1) = -36$

 3 Solve these equations.

    a $\dfrac{a - 2}{4} = 3$ **b** $\dfrac{b + 6}{2} = 5$ **c** $\dfrac{c + 9}{11} = 4$

    d $\dfrac{d - 4}{7} = 5$ **e** $\dfrac{e + 3}{8} = 4$ **f** $\dfrac{2f - 7}{3} = 6$

  4 Helmut's homework and solutions are shown below. Explain what is wrong with each solution. Find the correct answer, where necessary.

Questions   Solutions

    a $9x - 2 = 25$ **a** $9x - 2 = 25 + 2 = 3$

    b $3 + 4y = 19$ **b** $3 + 4y = 19$

    c $2(2x - 3) = 14$ $4y = 19 + 3 = 22$

                           $y = 5.5$

                    c $2(2x - 3) = 14$

                          $4x - 3 = 14 \Rightarrow 4x = 17$

                          $x = 4.25$

  5 The formula for the perimeter of a rectangle is given by $P = 2(a + b)$.

     a and b are different positive integers.

    a Write down all the possible pairs of values for a and b if $P = 16$.

    b Write down all the possible values for P if the area of the rectangle is 30.

    c Find the value of a if $P = 356$ and $b = 29$

    d Explain why the rectangle cannot have a perimeter of 35.

6 Solve these equations. Give the answers as mixed numbers.

    a $5(x - 2\frac{1}{3}) = 24$ **b** $4(x - 3\frac{3}{5}) = 19$ **c** $7(x + 2\frac{1}{2}) = 6$

7 Solve these equations. Write the answers as decimals. Do not use a calculator.

    a $3.6p + 2.8 = 17.56$ **b** $\dfrac{2q - 1.4}{3} = 2.8$ **c** $5(w + 3.2) = 29.5$

15.2 Equations with the variable on both sides

1 Solve these equations.

    a $5x = x + 20$ **b** $3x = x + 14$ **c** $8x = x + 35$

    d $4x = x + 18$ **e** $7x = x + 84$ **f** $9x = x - 16$

2 Solve these equations.

    a $5y = 9 + 2y$ **b** $10x = 20 + 6x$

    c $8u = 3u + 25$ **d** $7p = 3p + 32$

3 Solve these equations.

    a $6g + 5 = 2g + 13$ **b** $9i + 7 = 5i + 19$

    c $10h - 3 = 3h + 18$ **d** $8t - 15 = 6t + 3$

4 Solve these equations.

    a $8y = 6y - 10$ **b** $6k - 6 = 9k$ **c** $40 + 7j = 2j$

    d $5d + 9 = 3d + 3$ **e** $-4r = 3r + 21$ **f** $7n - 3 = 3n - 15$

5 a Find the area of this rectangle.

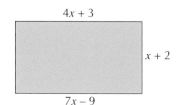

    b Find the area of this square.

6 Solve these equations.

    a $20 - 3x = x$ **b** $12 - 2c = 2c$

    c $4d = 30 - d$ **d** $5p = 14 - 2p$

7 Solve this equation. Leave your answer as a mixed number.

    $23 - 3\frac{1}{3}x = 14 + 1\frac{2}{3}x$

8 Solve these equations.

    a $6 + 2x = x - 5$ **b** $6 + x = 5x - 2$ **c** $5x + 2 = 6 - x$ **d** $2 + x = 5 - 6x$

  9 Simrath gets £8 pocket money per week. Ruby gets £10 pocket money per week.

Simrath spent £x of her pocket money on stamps.

    a Write an expression for the amount of money Simrath had after she bought her stamps.

    Ruby bought five times as many stamps as Simrath.

    b Write an expression for the amount of money Ruby had after she bought her stamps.

    Simrath and Ruby each had the same amount of money left after buying their stamps.

    c Write and solve an equation to find out how much a stamp cost.

    d How much money did Ruby have left after buying her stamps?

Brainteaser

Complete the crossword.

Across

1   $4(x - 35) = 13\,052$

3   $8x - 43\,210 = 3x - 14\,725$

7   $634 - 4x = 4000 - 10x$

8   $4x - 9301 = 600\,000 - 3x$

9   $9(x - 1234\,567) = 14\,010\,687$

13   $15(x - 9090) = 688\,110$

15   $1000 - x = 10x - 3532$

16   $23\,032 = 8(x + 4)$

17   $15x + 42\,167 = x + 97\,131$

Down

1   $32(x + 111) = 114\,176$

2   $3(x + 46) = 294\,579$

4   $12(x - 503) = 1404$

5   $9x - 222 = 5x + 29\,122$

6   $8x + 4\,404\,645 = 11x - 999\,999$

10   $6(x - 535) = 95\,664$

11   $761 + x = 3945 - x$

12   $10(x - 8255) = 10$

14   $10\,000 - 17x = 1567 - 8x$

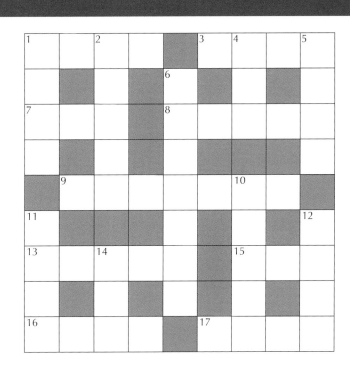

15.3 More complex equations

 Solve these equations.

a $5x + 2 = 20$ b $5(x + 2) = 20$ c $5(x + 2) = x + 20$ d $5(x - 2) = 20 - x$

 Solve these equations.

a $3(x + 2) = 2x + 8$ b $3(2g + 3) = 4g + 17$

c $5(s - 2) = 3(s + 4)$ d $8w - 12 = 3(3w - 1)$

 a   Copy and complete this table.

x	1	2	3	4	5
$4(x + 6)$					
$8(9 - x)$					

b   Use the table to solve the equation $4(x + 6) = 8(9 - x)$.
c   Solve the equation algebraically and check that you get the same answer.

 Solve the following equations. The answers to the equations are the whole numbers from 1 to 8. Each answer should be used once.

a $2(a + 6) = 3(a + 3)$ b $4(b - 1) = 2(b + 6)$ c $5(c + 2) = 3(c + 6)$

d $2(d + 2) = 4(d - 2)$ e $6(8 - e) = 12(e + 1)$ f $11(f + 3) = 22(f - 2)$

g $3(g + 3) = 2(7 - g)$ h $4(h - 3) = 8(h - 4)$

 Solve these equations.

a $6(x + 3) = 10(x + 1)$ b $2(x - 1) = 5(x - 7)$ c $2(13 - x) = 4(8 - x)$

d $3(x + 6) = 4(x + 3)$ e $8(x + 10) = 12(x + 8)$ f $5(x + 11) = 10(x + 7)$

g $9(x - 14) = 3(x - 2)$ h $3(x + 15) = 4(x + 13)$

 Each of the sides of a hexagon is $(y + 6)$ metres long.

Each of the sides of a pentagon is $(y + 8)$ metres long.

The hexagon and pentagon have the same perimeter.

a   Write expressions for the perimeter of each shape.
b   Solve an equation to find the value of y.
c   Work out the perimeter of the hexagon.

(PS) Pair up the equations with the same answers.

a $3(w + 6) = 9(w - 2)$ b $3(w + 3) = 6(w - 6)$ c $6(w + 2) = 2(w + 11)$

d $5(10 - w) = 10(w - 4)$ e $8(w + 7) = 12(w + 6)$ f $9(w + 12) = 6(w + 16)$

g $9(w - 1) = 3(w + 2)$ h $10(22 - w) = 2(w + 20)$

8 Solve these equations.

  a $\frac{2}{5}(w - 4) = 6$

  b $\frac{x - 5}{4} = x - 11$

  c $\frac{1}{2}(y + 9) = 2(y - 6)$

  d $\frac{3z + 2}{4} = z + 5$

  9 Ciara is c years old. She is six years older than her sister.

In two years' time she will be three times as old as her sister.

How old is Ciara?

15.4 Rearranging formulae

1 Rearrange each formula to make m the subject.

  a $r = m - 3$ **b** $r = 4(m - 3)$ **c** $r = 4m - 3$ **d** $r = \frac{1}{4}m - 3$

2 Rearrange each formula to make b the subject.

  a $a = b + c$ **b** $a = b - c$ **c** $a = bc$ **d** $a = \frac{b}{c}$

3 Rearrange each formula to make x the subject.

  a $y = x + 9$ **b** $y = 4x$ **c** $y = 5x - 1$

  d $y = 6(x + 5)$ **e** $y = \frac{1}{3}(x + 1)$ **f** $y + 3x = 8$

4 Rewrite each of the following formulae as indicated.

  a $A = 9p$ Make p the subject of the formula.

  b $y = x + 5$ Make x the subject of the formula.

  c $A + 5 = C$ Make A the subject of the formula.

  d $y = 8 + c$ Make c the subject of the formula.

  e $T = 4m$ Make m the subject of the formula.

5 The perimeter of a shape is given by the formula $P = a + 5$.

  a Find the value of P when $a = 4$ cm.
  b Make a the subject of the formula.
  c Calculate the value of a when $P = 37$ cm.

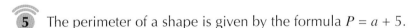 **6** The speed of a car is u mph. It accelerates to a speed of v mph.

Its final speed is given by the formula $v = u + 16$.

  a Calculate the final speed of the car if it accelerates from 20 mph.
  b Make u the subject of the formula.

 7 $A = a + 4$

  a Find A when $a = 5$.
  b Make a the subject of the formula.
  c $A = 18$. Use your formula to find a.

8 Here is a pentagon.

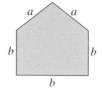

    a Show that the formula for the perimeter, p, is $p = 2a + 3b$.
    b Work out the value of p if $a = 5$ and $b = 6$.
    c Rearrange the formula to make a the subject.
    d Rearrange the formula to make b the subject.

 9 Here is a heptagon.

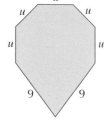

    a Show that the formula for the perimeter, P, is $P = 5u + 18$.
    b Rearrange the formula to make u the subject.
    c Find the value of u when the heptagon has a perimeter of 173.

 10 The formula $D = \frac{M}{V}$ is used in science.

    a Find the value of D when $M = 30$ and $V = 6$.
    b Rearrange the formula to make M the subject.
    c Find the value of M when $D = 12$ and $V = 4$.
    d Rearrange the formula to make V the subject.
    e Find the value of V when $M = 40$ and $D = 5$.

11 The area of a trapezium is given by the formula $A = (a + b)h$.

    a Show that the formula can be rearranged as $a = \frac{2A}{h} - b$
    b Rearrange the formula to make b the subject.

Brainteaser

Laurel has some budgerigars and some cages.

If six budgerigars are put in each cage, one budgerigar is left over.

If seven budgerigars are put in each cage, one cage is left over.

How many budgerigars and how many cages are there?

16 Comparing data

16.1 Grouped frequency tables

 1 The table shows the lengths (L metres) of 30 snakes.

Length of snake (L metres)	Frequency
$3.5 \leqslant L < 4$	2
$4 \leqslant L < 4.5$	4
$4.5 \leqslant L < 5$	7
$5 \leqslant L < 5.5$	9
$5.5 \leqslant L < 6$	5
$6 \leqslant L < 6.5$	3

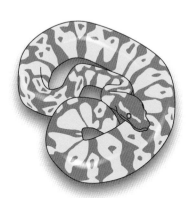

a One of the snakes is 4.5 metres long. Which class contains this length?
b How many snakes are shorter than 5 metres?
c How many snakes have a length of 5.5 metres or more?
d How many snakes are between 4 and 6 metres long?
e How many snakes could be exactly 5 metres long?

  2 Twenty people at a party play a memory game. They are shown 20 objects for a minute before they are removed. They then have to write down as many as they can remember. The table shows how they did:

Objects remembered	Number of people
1–5	0
6–10	2
11–15	12
16–20	6

a What was the modal class?
b How many people remembered more than half the objects?
c Could anyone have remembered them all?
d What percentage only remembered half or less?
e What is the minimum possible range?
f What is the maximum possible range?

3 The party game in question **2** was followed by a music quiz. They were played clips from 25 songs and had to write down what they were and who sang them. Their scores are shown below:

4, 5, 8, 8, 8, 11, 12, 12, 14, 15, 15, 16, 17, 18, 18, 20, 21, 21, 21, 23

a Put these results into a table like the one below:

Score	Tally	People
1–5	//	2
6–10		
11–15		
16–20		
21–25		

b What was the range of their scores?
c How many separate modes are there?
d What was the modal class?
e Are the modal scores inside the modal group? Explain what has happened here.
f What proportion of players got less than half correct?

 4 Twenty girls compare how much money they received for their birthdays, in pounds:

20, 25, 32, 35, 40, 45, 50, 50, 52, 55, 60, 60, 70, 70, 70, 75, 75, 80, 85, 90

Money (£)	Tally	Frequency
$0 < M \leq 20$		
$20 < M \leq 40$		
$40 < M \leq 60$		
$60 < M \leq 80$		
$80 < M \leq 100$		

a Copy and complete the frequency table of this data.
b Without knowing the actual data what could the maximum range be? What is the actual range?
c Why is the actual mode different to the modal class?
d What proportion of the girls received more than £50?

 5 The volumes (V cl) of liquid contained in 20 coconuts are shown below.

12.2	11.1	10.5	12.8	12.0	10.1	11.8	12.3	10.7	12.7
10.0	11.6	12.1	10.5	10.8	12.6	10.7	11.4	12.8	11.3

a Copy and complete the table.

Volume of liquid (V cl)	Tally	Number of coconuts
$10 \leqslant V < 10.5$		
$10.5 \leqslant V < 11$		
$11 \leqslant V < 11.5$		
$11.5 \leqslant V < 12$		
$12 \leqslant V < 12.5$		
$12.5 \leqslant V < 13$		

b What is the range of the volumes of liquid?

c What is the modal class?

d What proportion of coconuts contain less than 11 cl?

e What is the probability that a coconut chosen at random contains 12 cl or more?

16.2 Drawing frequency diagrams

1 Draw a bar chart for question **1** in section 16.1.

2 At a children's party the following balloon colours were used:

pink, red, pink, red, yellow, green, blue, red, yellow, pink, blue, green, pink, red, yellow, green, red, blue, yellow, red.

Draw a colourful bar chart to illustrate this.

FS **3** Shop prices for the same beach ball are as follows. Draw a bar chart to illustrate the data.

Price of beach ball (£P)	Number of shops
$2.80 < P \leqslant 3$	27
$3 < P \leqslant 3.20$	51
$3.20 < P \leqslant 3.40$	30
$3.40 < P \leqslant 3.60$	20
$3.60 < P \leqslant 3.80$	13
$3.80 < P \leqslant 4$	9

4 Amounts of coffee served in 500 ml cups from a survey of cafes are shown in the table below.

a Draw a bar chart for the data.

Amount of coffee (*V* ml)	Number of cups
$485 \leqslant V < 490$	3
$490 \leqslant V < 495$	7
$495 \leqslant V < 500$	13
$500 \leqslant V < 505$	16
$505 \leqslant V < 510$	11

b The table shows that some cafes are cheating their customers by giving short measures whilst others are being a little generous. Using percentages, show whether more cafes are giving too little or too much.

5 A patient's temperature was taken every hour to check on their wellbeing. 36.8 °C is the normal temperature for a human, and anywhere over 37.5 °C is classed as fever. The results are as shown:

9:00 am	36.9 °C
10:00 am	37.0 °C
11:00 am	37.3 °C
12:00 noon	37.6 °C
1:00 pm	37.8 °C
2:00 pm	37.9 °C
3:00 pm	37.9 °C
4:00 pm	37.7 °C
5:00 pm	37.6 °C
6:00 pm	37.4 °C
7:00 pm	37.2 °C
8:00 pm	37.0 °C
9:00 pm	36.8 °C

a Draw a line graph to show these changes in temperature.
b As accurately as possible, when would you say the fever started and finished?
c What could you draw on your graph to help you answer part **b**?
d Did the fever last for more or less than a quarter of a day (24 hours)?
e At about what time do you think the medication started to reduce the fever?

6 The table shows the average height of sweetcorn plants after being sprayed with different amounts of a new fertilizer.

Amount of fertilizer (A ml)	Height of plant (h m)
0	1.40
10	1.40
20	1.45
30	1.55
40	1.70
50	1.70
60	1.60
70	1.50
80	1.40
90	1.30
100	1.30
110	1.30

a Plot a graph. Use the following scales:

x-axis (amount of fertilizer): 1 cm to 10 ml

y-axis (height of plant): 2 cm to 0.1 m

b Estimate the level of fertilizer that would give plants a height of 1.5 m.

c Estimate the height of a plant sprayed with 35 ml of fertilizer.

d Which level of fertilizer would you advise the farmer to use? Explain your answer.

e At which levels did the fertilizer not improve growth?

16.3 Comparing range and averages

  1 **a** Find the mean and range of 21, 24, 25, 25, 26 and 29.
b Find the mean and range of 7.7, 7.3, 7.9, 7.1, 7.8, 7.2 and 7.5.
c Find the mean and range of 582, 534, 518, 566 and 550.
d Can you see a quick way of finding the mean for each of the questions above?

  2 **a** Write down two numbers with a range of 4 and a mean of 12.
b Why is there only one possible solution to part **a**?
c Write down three numbers with a mean of 10 and a range of 6.
d Why are there only three possible solutions to part **c**?
e What is crucial about the total of the three numbers in part **c**?

  3 **a** Find the mean and range of 2, 4, 6, 8 and 10.
b Add 5 to each number in part **a**. How are the mean and range affected?
c Double each number in part **a**. What happens to the mean and range now?
d Add x to each number in part **a**. Work out the new mean and range.

4 For each set of data, decide whether the range is a suitable measure of spread or not.

    a 30, 60, 100, 120, 150, 200   **b** 13, 45, 48, 52, 66

    c 5, 10, 15, 20, 25, 25, 25, 25, 25

  5 The weekly wages of four workers are £320, £290, £420 and £370.

    a Calculate the mean and range for the wages.

    b The employer hires a fifth worker. She wants the mean wage to be £340. What should she pay the fifth worker?

    c All workers receive a pay rise of £30 per week. How will this affect the mean and range? (Do not recalculate them.)

    d The following year the workers receive a 10% pay rise. How do you think the mean and range will be affected?

 6 QuickDrive and Ground Works are two companies that lay drives. The numbers of days each company takes to complete nine drives are shown below.

    QuickDrive   2, 5, 3, 3, 6, 2, 1, 8, 3
    Ground Works   3, 2, 3, 1, 4, 3, 2, 2, 1

    a Calculate the mode, mean and range for each company.

    b Comment on the differences between the averages.

    c Comment on the difference between the ranges.

  7 The table shows the mean and range for the average weekly rainfall (mm) in two holiday resorts.

	Larmidor	Tutu Island
Mean	6.5	5
Range	33	62

Explain the advantages of each island's climate using the mean and range.

Brainteaser

The table shows the number of goals scored by players at two different football clubs

Goals scored	Team A	Team B
0–4	16	18
5–9	10	9
10–14	5	4
15–19	3	4
20–24	1	3
25–29	0	2
30+	1	0

a Why might several players have scored no goals?

b Which team looks like it has one star player?

c Which team has the biggest squad?

d What proportion of both teams has scored   **i** over 10 goals   **ii** over 20 goals?

e Which team do you think is likely to have been most successful? Use figures to back up your answer.

16.4 Which average to use?

 1 For each set of data, do the following.

    i Calculate the given average.

    ii If the chosen average is unsuitable, give a reason why.

      a Mode   5, 5, 7, 9, 13, 13, 13

      b Mean   23, 25, 26, 29, 31, 36, 40

      c Median   11, 13, 15, 15, 16, 16, 16, 18, 18, 20, 80

      d Mean   0.3, 0.3, 2.3, 2.6, 2.9, 3.0

      e Median   120, 125, 140, 170, 1200, 1300, 2000

  2 Two shops offer the following sizes of dresses.

Periwinkle   12, 14, 16, 18, 30

Jenny's   10, 12, 14, 16, 18, 20, 22

    a Which shop has the greater range of sizes?

    b Do you think this is a suitable indication of sizes available? Explain your answer.

    c Would any of the averages be more helpful to potential customers and, if so, which one?

  3 Judges awarded the following points to competitors in a surfing competition.

23, 6, 47, 29, 41, 17, 38, 21, 30, 24, 40, 7, 27, 18, 20, 6, 26, 35, 28

    a Calculate the mode, median and mean.

    b Which averages are suitable?

    c Which averages are unsuitable? Explain your answer.

    d Copy and complete the tally chart.

Points awarded, p	Tally	Frequency
$0 \leqslant p < 10$		
$10 \leqslant p < 20$		
. . .		

    e State the modal class. Is this a suitable average?

  4 The midday temperature, in degrees Celsius, for 10 days during June in London were as follows:

17 18 20 24 26 28 23 17 17 16

    a Calculate the mode, median and mean for these figures.

    b Which do you think is the better average to use and why?

5 The table shows the weights of fish (in grams) caught by three anglers in a competition.

Jerry	230	100	380	520	450			
Adita	230	400	280	320	250			
Marion	100	130	630	200	70	180	200	90

a Calculate the mean and range for each of the anglers.
b Who was the most consistent? Explain your answer.
c Who performed the best overall? Explain your answer.
d Which angler would you choose to enter a competition that offered prizes for the heaviest fish caught? Explain your answer.

  6 Which average would be the best representative of these distributions – mode, median or mean? Do the calculations and explain your reasoning.

a 4 6 6 3 6 7 5 3
b 3 3 3 5 7 7 28
c 10 10 10 30 35 40
d 1.5 2.8 3.2 6.9 7.8 7.8 8.1
e $\frac{1}{4}$ $\frac{1}{4}$ $\frac{1}{2}$ $\frac{3}{4}$ $1\frac{1}{4}$ $1\frac{3}{4}$ $2\frac{1}{4}$

Brainteaser

A cricket team records the number of times certain scores were achieved during a tour abroad and their results are shown in the table below.

Runs scored	Number of times scored (frequency)
0–24	10
25–49	8
50–74	9
75–99	6
100–149	4
150–199	2
200+	1

a How many innings were played in total?
b Over 100 runs were scored seven times. Does this necessarily mean seven different players scored a century?
c What percentage of the innings did not reach 50 runs?
d The modal class is 0–24. Do you think this is a good average to use to represent how the team did overall?
e What is the minimum and maximum number of runs the team could have scored?
f Which class do you think contains   **i** the median score   **ii** the mean score?
g What extra information do you need to find an accurate average?
h Can you see a way of finding an approximate mean from the data supplied?

Notes

Notes